Satellite Orbits and Communications

Edition 1

To the reader:
Thank you for joining us in this collaborative journey to make aerospace education more accessible. We made this book with the primary goal of creating a friendly guide for understanding the physics of satellite systems and the methods for sending data to and from space. This is the first edition of our book, so there are bound to be ways we can improve. If you have any thoughts or comments, please send them to us at radiant.edtech@gmail.com. Additionally, if you are an educator and would like the solutions manual, please email us to request a copy. Without further ado, time to blast off!

**Thank you to our reviewers
who helped make this textbook happen:**
Sreenidhi Chalimadugu, Neel Dhulipala, Angela Huang, Utsav Gupta, Alex Mineeva, Satchel Sevenau, Ethan Chen, Kat Canavan, George Hansel, Emily Calandrelli, Mia Chevere, Camden Droz, and Audrey Renaud.

Contents

Chapter 1

Preface

As its name suggests, *satellite communications* focuses on interactions to and from a *satellite*, which is an object that orbits a body in space. This field is broad and covers orbital mechanics, wireless communications, electromagnetic propagation, and hardware design. In this text, we dive into these topics while also providing relevant historical and social implications of this technical industry.

Throughout this book, our friends Sat, Squit, and Doggo will orbit by your side, ready to support you in your learning adventures! Sat is a mischievous bird that loves trivia and exercises. Whenever you see Sat, an example problem is sure to follow!

Example with Sat!

Skwaa! Nice to meet you, as we say in bird speak. My name is Sat, and I'll be the one who challenges your noggin' to make sure you understand the concepts we introduce.

Doggo is your loyal companion, who will provide helpful tips, and who will make sure you don't get lost along the way.

Doggo Help

Woof! My name is Doggo. I love treats, belly rubs, and helping people! Look out for me for my friendly tips.

Squit is our octopod historian who loves nothing more than a fun fact! Squit will always have something fascinating to share.

Table 1.1: Table of Constants

Constant	Variable	Value	Unit
Earth radius	R_E	6378	km
Earth mass	M_E	5.974×10^{24}	kg
Constant of gravitation	G	6.67×10^{-11}	Nm^2/kg^2
Earth gravitational parameter	μ	398,600	km^3/s^2
Speed of light	c	3.0×10^8	m/s
Boltzmann's constant	k	1.381×10^{-23}	$\text{m}^2\text{kg s}^{-2}\text{K}^{-1}$
Geostationary altitude	h_{GEO}	35,786	km

History with Squit!

Sploof! Howdy, my name is Squit: your one stop shop for history. Some may think of history as just boring dead people, but history is a living thing that we are all continuously making. Look out for me to learn about the past, present, and future!

We actually have one more member of our team: **you**. All of us, including Sat, Squit, and Doggo, want to support your learning and see you succeed. If you ever feel overwhelmed, don't panic, just take a step back and revisit what you've learned. Space can feel hard to comprehend at first, but we'll be here with you every step of the way. You can do this!

1.1   Common Constants

Throughout this text, we'll encounter a variety of repeated numbers that can be difficult to remember. Understanding these constants is key, as they are fundamental to grasp the principles of how satellites operate! Table 1.1 summarizes a collection of constants that we'll need to keep on hand to navigate the complexities of orbital mechanics and satellite communications.

Together, we'll explore how these constants are not just abstract figures but integral to the design of satellite systems.

1.2 Common Variables

We will also encounter equations that use variables denoted with symbols. For reference, Table 1.2 provides definitions of the most common variables in the book.

Table 1.2: Table of Common Variables

Variable	Meaning	Description
d	distance	Measurement of separation between two points
r or R	radius	Distance between center and edge of an arbitrary body
h	altitude	Normal distance between the surface of the Earth and satellite, or some point in space
r_a	apogee	Distance between the orbited body (e.g. Earth) and the farthest point from it on the orbit
r_p	perigee	Distance between the orbited body (e.g. Earth) and the closest point to it on the orbit
a	semi-major axis	Radius of largest axis on ellipse; distance between center and farthest point on an orbit
e	eccentricity	Ratio of the distance between both focal points and semi-major axis of an orbit
i	inclination	Angle between the satellite's orbit and the equatorial plane
Ω	RAAN	Right Ascension of Ascending Node: angle on the equatorial plane between the Prime Meridian and the ascending node
ω	argument of perigee	Angle between the ascending node and perigee of the orbital plane
ν	true anomaly	Angle on the orbital plane between the satellite and perigee

E	eccentric anomaly	Angle on the orbital plane between the perigee and the point on an auxiliary circle circumscribing the satellite's orbit
M_e	mean anomaly	Position, in radians, representing the fraction of an elliptical orbit since the satellite passed through perigee
T	period	Amount of time taken for one object to complete an orbit around another object
λ	wavelength	Distance between two peaks (or troughs) of a wave
f	frequency	Number of cycles or repetitions of a wave per unit of time
T_x	transmitter	When attached as a subscript, refers to the property of a *transmitter*
R_x	receiver	When attached as a subscript, refers to the property of a *receiver*
P	power	Amount of power in a signal transmitted or received by an antenna
G	gain	Ratio between the power radiated from an antenna and the theoretical power radiated from an isotropic antenna

Some of the variables in this table may have multiple meanings. For example, while h usually refers to altitude in this text, h can also represent specific angular momentum. This is precisely the case for Chapter 3 Section 3.4! Additionally, it may seem that d and R, representing distance, are used interchangeably throughout the book. Certain equations will use d if the distance is more of an arbitrary measurement, while other equations will use R if that distance is a radius from a central point. However, in practice, this may not always be the case! We'll make sure to clarify the definition of each variable whenever we use them.

Part I

Satellite Dynamics

Chapter 2

Orbit Types and Elements

2.1 Types of Orbits

When you look up into the sky, thousands of satellites fly overhead, even though you might not see them. As shown in Figure 2.1, most satellites range in size from objects that can fit in your hand to objects as big as a school bus—and some, like the International Space Station (ISS), can be as large as a building! The main structure of a satellite is often referred to as the "bus". We'll stick mostly to orbits and communication theory in this book, but it's important to know that numerous subsystems, like the power, thermal, and structures subsystems exist, and must all interface seamlessly to execute the mission successfully.

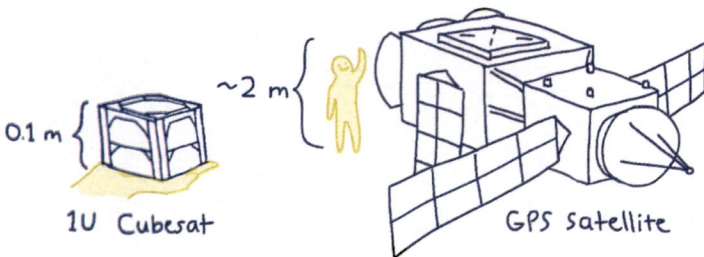

Figure 2.1: Different Satellite Bus Sizes

To start, we are going to learn about *orbital mechanics*. Orbits are curved paths an object follows around a celestial body. For example, the Moon is a natural satellite that orbits around the Earth. Generally, Earth-centered orbits are categorized as one of three types: Low

Earth Orbit (LEO), Medium Earth Orbit (MEO), and Geostationary Orbit (GEO). Each type of orbit is generally considered to exist at the following altitudes:

- LEO: 0 - 2,000 km

- MEO: 2,000 - 35,786 km

- GEO: 35,786 km

Figure 2.2 outlines these three primary orbits. The acronym NGSO (Non-Geostationary Orbit) represents all satellite systems that are not orbiting at an altitude of 35,786 km, including all LEO and MEO satellites.

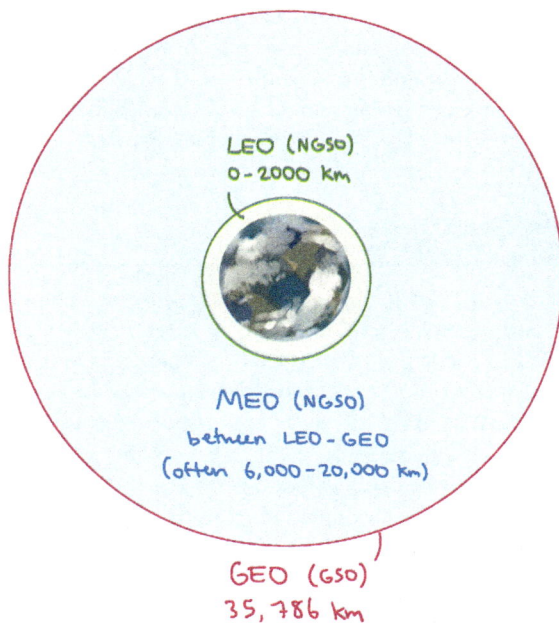

Figure 2.2: LEO, MEO, and GEO (not to scale)

You may have noticed that LEO and MEO consist of altitude ranges, while GEO exists at a specific height of 35,786 km above the Earth's surface. Geostationary orbits are fixed at the equator and are a specific sub-category of geosynchronous orbit, which can orbit at any inclination. Figure 2.3 shows the distinction.

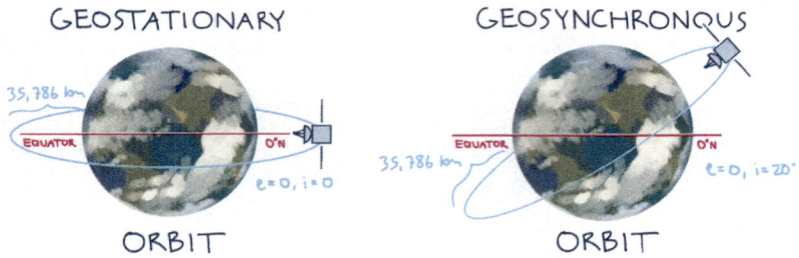

Figure 2.3: Geostationary vs. Geosynchronous Orbits

Geostationary orbits are positioned at an altitude where the satellite's orbital period matches the rotational period of the Earth (23 hours, 56 minutes, and 4 seconds). This alignment allows geostationary satellites to remain stationary over a single point on the Earth's surface. This length of time is defined as a **sidereal day**. As a result, a geostationary satellite placed directly over a location, like Nairobi, Kenya, will always stay fixed directly over Nairobi because the satellite is orbiting in tandem with the Earth's rotation.

History with Squit!

Sploof! In our everyday lives, when we say there are 24 hours in a day, we are referring to a *solar day*, which is the time it takes for the Sun to return to the same position overhead. A *sidereal day* is used by astronomers because it is based on "Earth's rate of rotation measured relative to the fixed stars," or inertial space [6]. Before the 1970s, sidereal time was measured using special telescopes that faced upwards to track and time the passage of stars [6]!

Unfortunately, the distinction between geostationary and geosynchronous satellites can be particularly confusing, as both GEO and GSO acronyms are used interchangeably to describe geostationary and geosynchronous orbits. It is essential to note that satellites in geostationary orbit are located directly over the Earth's equator so that the satellite's longitudinal AND latitudinal location over the Earth stays fixed. On the other hand, geosynchronous orbits can be inclined and so will traverse a latitudinal range in their path.

There are also **Highly Elliptical Orbits (HEO)**, which often extend past GEO in **apogee**, the point in its orbit farthest from Earth, and reach LEO altitudes in **perigee**, the point in its orbit closest to Earth (we'll learn more about these terms soon). Check out Figure 2.4 for a depiction of a HEO spacecraft.

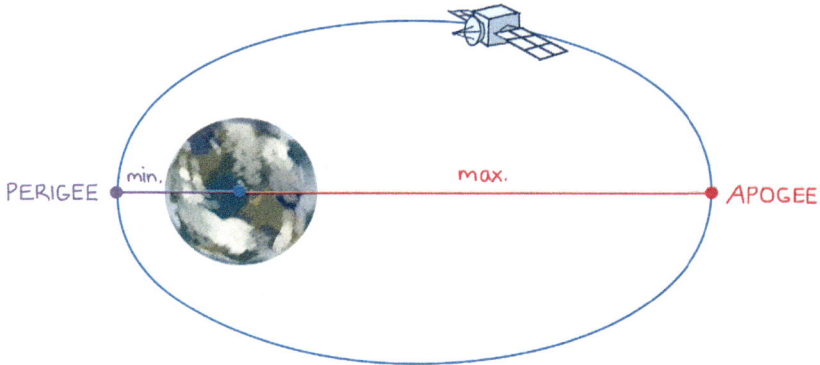

Figure 2.4: Highly Elliptical Orbit (HEO)

Highly elliptical orbit are egg-shaped and are designed to dwell over a certain region of Earth. The Molniya orbit is an example of a HEO that the Soviet Union developed to maximize the amount of time a given spacecraft would hover over high-latitude Soviet territory.

History with Squit!

Molniya (Russian: Молния) comes from the Russian word for lightning. But according to Squit's calculations, satellites in these orbits aren't very much like lightning since they travel at only a fraction of the speed of light!

The last type of orbit we will cover is the **Sun-Synchronous Orbit (SSO)**, illustrated in Figure 2.5. An SSO is a highly inclined orbit that typically has an altitude between 600 and 800 km and is designed such that the satellite always passes over the same longitude at the same angle to the sun. To do this, the satellite orbits ~1 degree eastward around the Earth daily, at the same rate that the Earth revolves around the Sun. Wow, that sure feels complicated! The main takeaway is that satellites in this orbit experience minimal variations in shadows and lighting, making an SSO orbit ideal for imaging and weather satellites

[19]. These orbits also place the satellite in constant sunlight, ensuring the solar panels on the satellite are continuously illuminated.

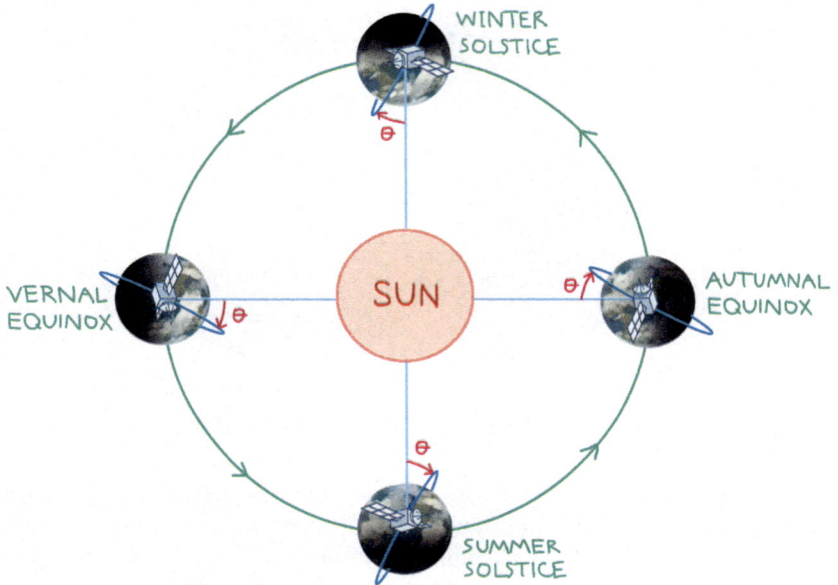

Figure 2.5: Sun-Synchronous Orbit (SSO)

History with Squit!

Note that the Greek suffix *-gee* means Earth and that we use apogee and perigee when Earth is the reference point. The Greek suffix *-apsis* generally refers to either of the two extreme points, known as apoapsis and periapsis. When the Sun is the reference point, the terms become perihelion and aphelion.

When the Earth is in the center of a circular orbit, the radius of apogee, r_a, and the radius of perigee, r_p, are equal. Therefore, the semi-major axis for a circular orbit equals the radius of the orbit. In an elliptical orbit, which is elongated and oval-shaped, the semi-major axis represents the average distance between the center of the Earth, one **focus** of the ellipse, and the satellite.

Remember to include the Earth's radius when defining altitude! In other words, **altitude** is the sum of the Earth's radius and the orbit height above the Earth's surface.

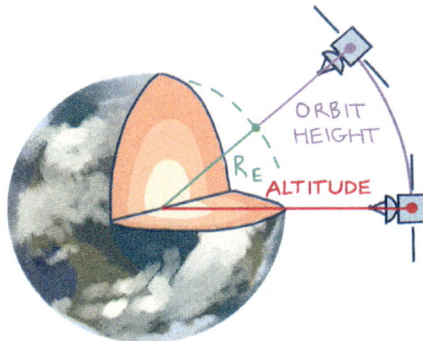

Figure 2.7: Altitude

In this text, we use a constant value of 6378 km for the Earth's radius R_E. More sophisticated calculations account for the Earth's oblateness, which implies that the radius from the center of the Earth to its surface is variable. In fact, the radius of the Earth is more prominent around the equator, with a maximum value of 6378 km, and is shorter at the poles, with a minimum distance of 6357 km [33].

Before we move on to our second orbital element, let's work through a sample calculation with Sat.

Example with Sat!

A satellite in LEO has an altitude of 1000 km at perigee and 1500 km at apogee. What is the satellite's radius of perigee? What is the semi-major axis?

Solution:

If you said 7378 km for the radius of perigee, then you're correct! The radius of perigee is found by summing the altitude at perigee, 1000 km, with the radius of the Earth, R_E.

For the semi-major axis, remember to add the radius of the Earth to both altitudes, then take the sum and divide by 2. This is the same as taking the average, but you must add R_E to both altitudes.

$$r_a = 1500 \text{ km} + 6378 \text{ km} = 7878 \text{ km} \tag{2.2}$$

$$r_p = 1000 \text{ km} + 6378 \text{ km} = 7378 \text{ km} \tag{2.3}$$

$$a = \frac{r_a + r_p}{2} = \frac{7878 \text{ km} + 7378 \text{ km}}{2} \tag{2.4}$$

Did you compute a value of 7628 km for your semi-major axis? If so, take a moment to pat yourself on the back. You did it!

The second orbital element, **eccentricity**, e, describes the shape of the orbit. Theoretically speaking, eccentricity describes the shape of a conic section, or the cross-section of a cone. An orbit's eccentricity, e, is the ratio of the distance between the two foci and the length of the major axis. It turns out that the distance between the two focus points is equal to the difference in the radius at apogee and the radius at perigee.

$$e = \frac{r_a - r_p}{r_a + r_p} \text{ [unitless]} \tag{2.5}$$

There are four primary conic sections, illustrated in Figure 2.8, that each corresponds to a specific value or range of eccentricity.

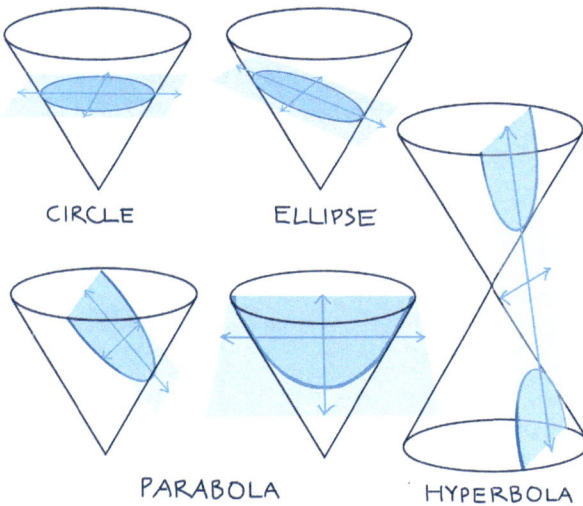

CIRCLE ELLIPSE

PARABOLA HYPERBOLA

Figure 2.8: Conical Sections

Circular and elliptical orbits are considered closed-loop, whereas hyperbolic orbits are considered open-loop. An object in a **closed-loop orbit** completes a loop around the celestial body they are orbiting. Objects in **open-loop orbits** have an eccentricity greater than one, meaning that any satellite with this orbit will escape the gravitational influence of its parent celestial body. Open-loop orbits are often used for interplanetary missions! Interestingly, parabolic orbits have an eccentricity of one and exhibit characteristics of both open- and closed-loop orbits.

- $e = 0$: Circular orbit are orbits whose apogee is equal to the perigee.

- $0 < e < 1$: Elliptical orbit are orbits whose shape is more akin to a surfboard.

- $e = 1$: When a satellite traveling along a parabolic orbit comes close to a celestial orbit, it acts like an elliptical orbit. But, when the satellite travels farther away into space, the gravitational effects of the parent celestial body diminish. This makes the orbit more hyperbolic, but the satellite will eventually return, acting like a pendulum, instead of escaping into space.

- $e > 1$: Hyperbolic orbits occur when a satellite exceeds the velocity necessary to escape the gravitational pull of the body the satellite is orbiting. Therefore, hyperbolic trajectories do not orbit the planet but rather escape the planet's gravitational field.

In summary, we can use eccentricity to classify an orbit as either: circular, elliptical, parabolic, or hyperbolic, as shown in Figure 2.9.

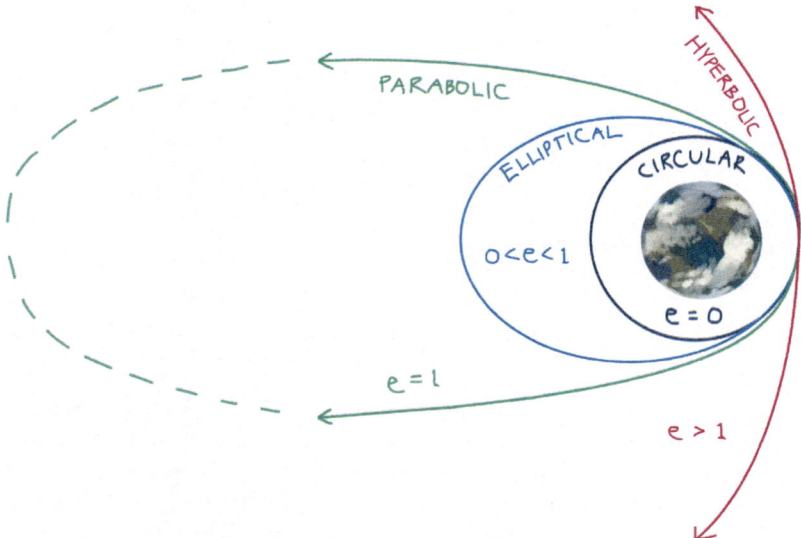

Figure 2.9: Four Orbit Types - Conic Sections

Figure 2.10 provides another way to visualize eccentricity as it increases from e = 0.01 to 0.99.

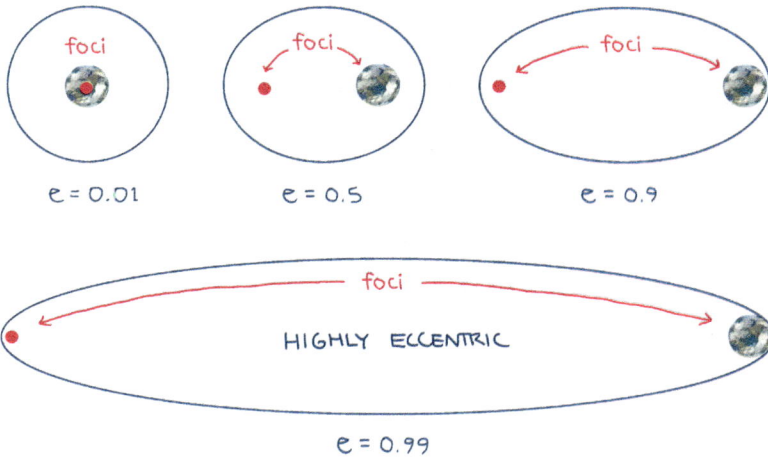

Figure 2.10: Eccentricity

Example with Sat!

What is the eccentricity of the LEO satellite in our previous example? Recall that its apogee is 1500 km and its perigee is 1000 km.

Solution:
Using the equation for eccentricity, we compute

$$e = \frac{r_a - r_p}{r_a + r_p} = \frac{1500 \text{ km} - 1000 \text{ km}}{1500 \text{ km} + 1000 \text{ km}} = \frac{500 \text{ km}}{2500 \text{ km}} = 0.2$$

Is this computation correct? NO! We must NOT forget to include the Earth's radius in our radius equations!!!

$$e = \frac{r_a - r_p}{r_a + r_p} = \frac{7878 \text{ km} - 7378 \text{ km}}{7878 \text{ km} + 7378 \text{ km}} = \frac{500 \text{ km}}{15256 \text{ km}} = 0.03$$

The third element, **inclination**, i, is the angle between the orbital and the equatorial plane, as illustrated in Figure 2.11.

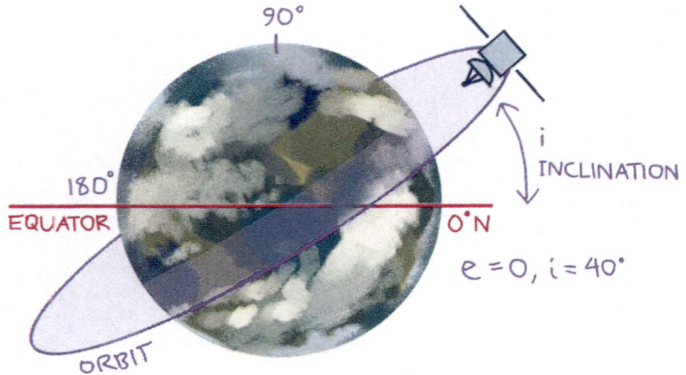

Figure 2.11: Inclination

Inclination ranges from 0 to 180 degrees, and determines the latitudes a satellite's ground track will cover. However, depending on the field of view of the instrument onboard, the satellite may be able to image or communicate with a wider range of latitudes. An inclination of 0 degrees defines an equatorial orbit. As you know by now, all geostationary satellites have an inclination of zero degrees. The next inclination of note is at 63.4 degrees, which is the critical inclination at which a satellite experiences "zero apogee drift", meaning that its farthest point away from the Earth does not vary! Lastly, a satellite orbiting at an inclination of 90 degrees is in a polar orbit and covers the Earth's North and South Poles.

We utilize inclination to define whether an orbit is in **prograde** or **retrograde**. Satellites in orbits with inclinations less than 90 degrees are in prograde, and appear to rotate along the same direction as the Earth's rotation (eastward or counter-clockwise). Satellites with inclinations greater than 90 degrees are in retrograde, and appear to rotate in the opposite direction as the Earth's rotation (westward or clockwise).

Doggo Help

Whoa! Great job so far! A helpful way to remember a prograde versus retrograde orbit is to use the right-hand rule. Try curling your right paw, or fingers, in the direction of the orbit. If your thumb pops up, then the orbit is prograde, and if your thumb turns down, then the orbit is retrograde. Doggo wishes he had opposable thumbs so he could also do cool memory tricks.

Table 2.1 summarizes the different types of orbits with characteristic orbital elements and examples of each!

Table 2.1: Types of Orbits

Orbit Type	Eccentricity	Inclination	Example
LEO $< 2,000$ km	< 1	All	Iridium OneWeb Starlink
MEO $2,000 - 35,786$ km	< 1	All	Mangata O3b mPOWER
GEO $35,786$ km	0	$0°$	Intelsat Echostar Viasat
HEO $> 35,786$ km apogee < 2000 km perigee	~ 0.75	$50 - 70°$	Molniya Tundra
SSO $600 - 800$ km	0	$\sim 98°$	Smallsats

2.3   Problem Set 1: Orbit Types and The First Three Orbital Elements

Problem	Topic	Points
1	Orbital Element Schematic	3
2	Perihelion and Aphelion	3
3	Orbital Parameters of 500 x 4000 km orbit	3
4	Orbital Parameters for a OneWeb Satellite	3
5	Altitude of GEO Satellite	2
6	Concept Map	3
Total:		17

Exercise 2.1

(**3 points**) Close your notes and sketch an elliptical orbit. Label the apogee, perigee, semimajor axis, eccentricity, and inclination.

1. Next, write the name, symbol and definition of each orbital element. Can you label any of the orbital elements on your sketch?

2. After you've given it your best attempt, refer back to figures in this section. How does the schematic you drew compare to the figures in this book?

3. When sketching and labeling your orbit, which elements were the first you remembered, which ones were more difficult to recall?

Exercise 2.2

(**3 points**) Sketch an elliptical orbit with the Sun at one of the foci. Label the aphelion (the location in the orbit that is furthest from the Sun) and perihelion (the location in the orbit that is closest to the Sun). Please also label the radius to apoapsis, r_a, and the radius to periapsis, r_p.

Exercise 2.3

(3 points) A satellite is in an elliptical orbit around Earth with a perigee of 500 km and an apogee of 4000 km. What is the radius of perigee, radius of apogee, semi-major axis, and eccentricity?

Exercise 2.4

(3 points) Compute the radius of perigee, radius of apogee, semi-major axis, eccentricity, and orbital period of a OneWeb satellite in LEO and compare the values to an SES O3b satellite in MEO. Here are some values you will need for computing the quantities listed above for OneWeb and SES O3b.

	OneWeb	**SES O3b**
Altitude	1200 km	8062 km
Apogee	1200 km	8062 km
Perigee	1200 km	8062 km
Inclination	87.9°	70.0°

Hint: While we have yet to discuss orbital period, T, here's a head start:

$$T = \frac{2\pi}{\sqrt{\mu}} a^{\frac{3}{2}}$$

where μ is Earth's gravitational parameter, and is equal to $398,600 \text{ km}^3/\text{s}^2$.

1. Are these orbits in prograde or retrograde?

2. Which system has the shortest orbital period? Why? Be sure to provide your answer in units that are meaningful. For example, 6,000 seconds is likely better converted to minutes or hours!

3. How do the eccentricities of these orbits compare? Are they elliptical or circular?

Exercise 2.5

(**2 points**) Now that you know how to calculate the orbital period for a satellite given characteristics such as its altitude, apogee, and perigee, we will calculate the altitude of a geostationary satellite given its orbital period! Pretend you don't know the altitude of a GEO orbit already!

1. Given the equation for the orbital period, T, in the previous exercise, solve for a. In this part of the exercise, do not plug in any numbers. Leave your equation in terms of variables.

2. Next, solve for a numerically. Hint: We know that for a satellite to be geo*stationary*, it must have an orbital period of the length of a sidereal day, which is 23 hours and 56 minutes.

3. Assuming the orbit is circular, what is the altitude of a satellite in GEO? How does it compare to our definition of a GEO satellite?

Exercise 2.6

(**3 points**) Construct a concept map and a separate equations sheet for material in the textbook thus far. Keep this handy, as we will revisit it throughout the text. Add equations, symbols, and sketches to your concept map.

2.4 Classical Orbital Elements Continued

We will now spend time learning about the remaining three orbital elements. First, we will discuss **Right Ascension of Ascending Node (RAAN)**, which is often represented with a capital Greek omega, Ω. RAAN is a degree measurement that determines the pivot of the satellite orbit around the Earth. Let's use Figure 2.12 to break down exactly what we need to find the RAAN.

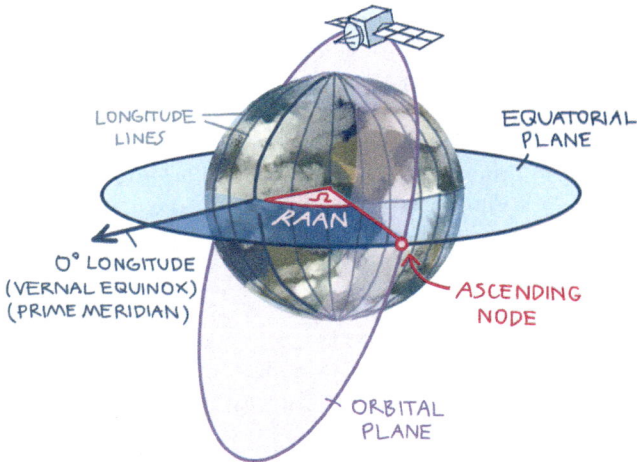

Figure 2.12: Right Ascension of the Ascending Node (RAAN)

Figure 2.12 may seem intimidating at first, but you have already seen most of the terms depicted. Let's orient ourselves! We first notice that this figure has two planes. The first one is the equatorial plane; imagine a giant circle cutting the Earth in half at the equator. The other plane is the satellite's orbit around the Earth. Since the satellite is ascending above the equator, the **ascending node**, is the location at which the satellite crosses the equatorial plane from south to north. Once we have the ascending node, we find the longitude of that point where the satellite crosses the equator. Since longitude is in degrees, that point is our RAAN! For this reason, the RAAN is also commonly known as the "longitude of ascending node."

Doggo Help

> *Bark!* Remember that longitude is represented by lines that cross the globe up and down, whereas latitude crisscrosses the globe from left to right! An easy way to remember this is "lat lay flat"! As a result, when we define longitude, we use east and west, and when we define latitude, we use north and south.

Before we move to our next element, there are a few more terms to cover. **Vernal equinox** is the point at which the Sun appears to cross the Earth's equator on its way north on the first day of spring each year [7]. This line is also known as the Prime Meridian and is where Greenwich Mean Time (GMT) is located. The angle between vernal equinox and the ascending node of our satellite is our RAAN since it is how far our orbit is pivoted from zero longitude!

History with Squit!

> *Darn tootin'!* How did both time and position end up centered in the same place? Well, in the 1760s, Astronomers at the Greenwich Observatory were the first to relate mechanical time on a clock to the activity of the Sun [1]. Before this, time was a local concept, unique to every town and village. Standardizing time and location would bring everyone on to the same page! But many countries disagreed about where the prime meridian should start! China made maps with the prime meridian running through Beijing, France made maps through Paris, and so on. It wasn't until 1884 that members from twenty-five countries chose the Greenwich Observatory as the international standard [25].

If RAAN determines how an orbit is pivoted around the equator, then the **argument of perigee**, ω, determines the swivel of the orbit around the Earth, as illustrated in Figure 2.13.

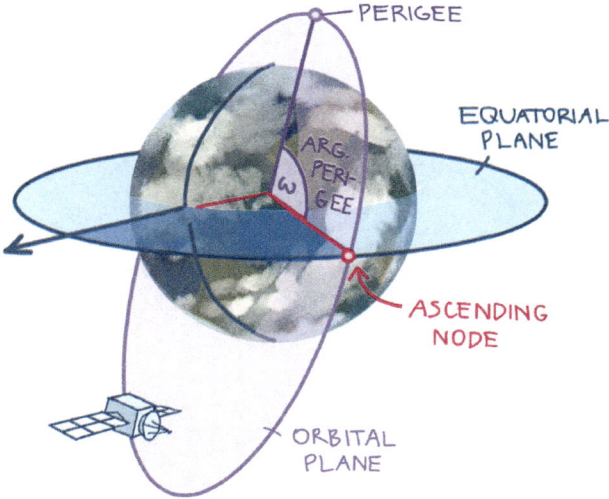

Figure 2.13: Argument of perigee

Imagine drawing a line from the center of the Earth to perigee. The argument of perigee, ω, is the angle between that line and the line extended out to the ascending node. Even though they may look similar, remember not to confuse argument of perigee with inclination. Inclination determines the latitudes a satellite will cover, whereas argument of perigee determines the closest and farthest point of an orbit.

We also want an idea of where our satellite is located along its orbit at any point in time. This is calculated based on a satellite's relative position to perigee. Let's start by drawing a diagram like Figure 2.14. Take that line from the center of the Earth out to the perigee. Now, draw a second line from the center of the Earth out to the satellite. The angle between these two lines is the **true anomaly**, ν.

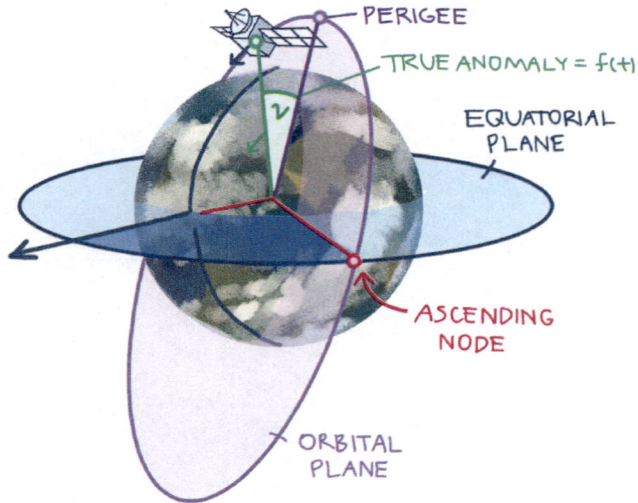

Figure 2.14: True Anomaly

As you may have discovered, the other elements are fixed relative to time, whereas true anomaly varies as a function of time. Satellites do not travel along non-circular orbits at a constant speed. Therefore, the true anomaly doesn't change at a constant pace either. To account for this, we calculate the **mean motion**, n, which is the average angular speed of the satellite in its orbit. To calculate mean motion, assume your orbit is circular and take the square root of Earth's gravitational parameter, μ, divided by the cube of the semi-major axis, a^3:

$$n = \sqrt{\frac{\mu}{a^3}} \; [\text{rad/s}] \tag{2.6}$$

We will learn more about μ in the next chapter when we take a closer look at Newton's Laws.

Example with Sat!

Kakaw! All of these classical orbit elements are making me feel like I have a bird brain! How about we try to draw an orbit that meets the following specifications:

- Elliptical orbit

- Inclination of 45°

- RAAN of 0°

- Argument of perigee of 90°

Use Figure 2.15 as a helpful guide if you forget how to define RAAN, and argument of perigee.

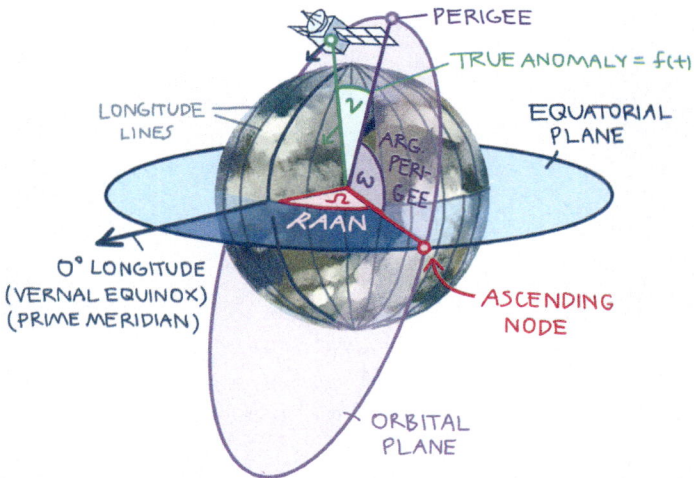

Figure 2.15: RAAN, Argument of Perigee, and True Anomaly

Example with Sat!

Solution:
SKWAH! How did you do?

Step 1. Identify the equator and target longitude.
Since our RAAN is 0°, our orbit crosses the equator at 0°.

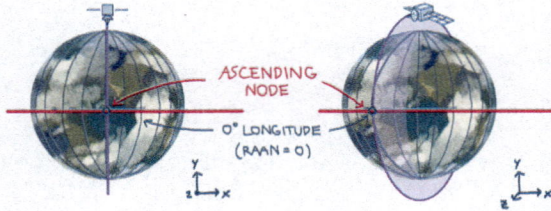

Step 2. Draw an initial orbit based on the inclination.

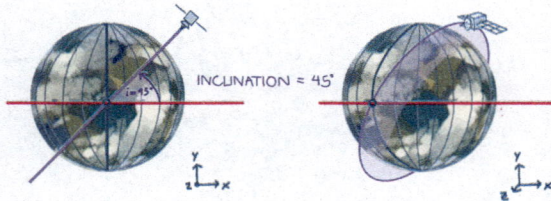

Step 3. Define the orbit's perigee (give it a nonzero eccentricity) and pivot it to 90° from the ascending node based on the argument of perigee.

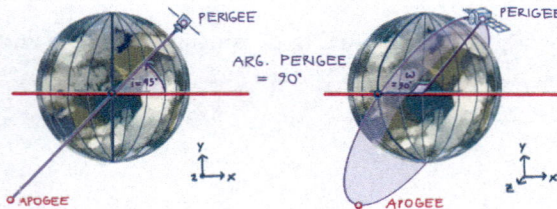

2.5 Mean Anomaly

The **mean anomaly**, M_e is based on the mean motion, n, of the satellite, and is the position in radians of a fictitious body moving around the ellipse at a constant angular speed. Mean motion, $n = \frac{2\pi}{T}$, is constant and, as we just saw, is equal to $\sqrt{\frac{\mu}{a^3}}$. We define the mean anomaly relative to the argument of perigee, ω, and, specifically, the time, τ, when our orbiting body is at the argument of perigee. At time, t, the mean anomaly is:

$$M_e = n(t - \tau) \tag{2.7}$$

$$= \frac{2\pi}{T}(t - \tau) \tag{2.8}$$

$$= \sqrt{\frac{\mu}{a^3}}(t - \tau) \tag{2.9}$$

To relate the true anomaly with the mean anomaly, we use the concept of eccentric anomaly, denoted as E. The **eccentric anomaly** is defined as the angle from the center of the orbit to the point that creates a right angle with the apse line, the line connecting the apogee and perigee, on a theoretical circle that includes the orbit. This theoretical circular orbit is used to simplify the calculation of the satellite's position.

These terms are easily and often confused. In summary, mean anomaly is the angle measured from the center of a circle circumscribing your elliptical orbit to a point along the circular orbit. The true anomaly is the angle from the apse line, the vector from the focus to periapsis, to the point in your elliptical orbit. Lastly, the eccentric anomaly is the angle measured from the center of the circle circumscribing the elliptical orbit to a point along the circular orbit that is dropped perpendicularly to the location of your object in the elliptical orbit.

The eccentric and mean anomaly of an orbit are provided in Figure 2.16.

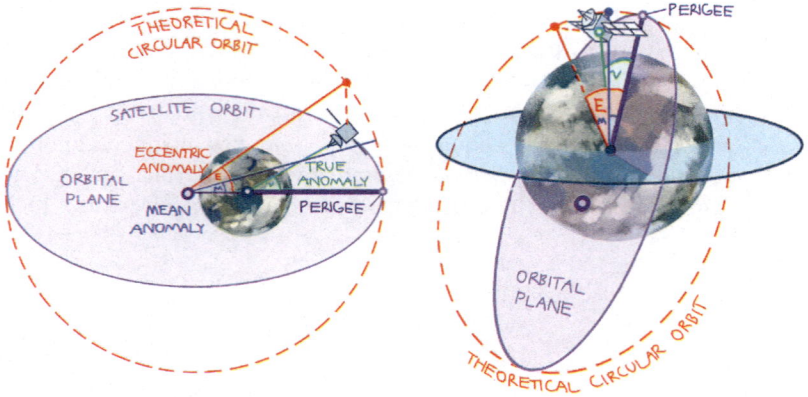

Figure 2.16: Eccentricity and Mean Anomaly of an Orbit

Kepler introduced what is known as Kepler's equation, where M_e is the mean anomaly of an elliptical orbit. For the math lovers out there, it can be helpful to remember that the mean anomaly monotonically increases as true anomaly increases.

$$M_e = E - e \sin E \ [\text{rad/s}] \tag{2.10}$$

One issue with Kepler's equation is that it's impossible to solve directly! We aren't going to dive into them here, but it will be useful to note two equations that come from the derivation of Kepler's Equation:

$$\cos \theta = \frac{e - \cos E}{e \cos E - 1} \tag{2.11}$$

$$E = 2 \arctan \left(\sqrt{\frac{1-e}{1+e}} \tan \frac{\theta}{2} \right) \ [\text{rad}] \tag{2.12}$$

In these equations, θ represents the true anomaly of the orbit. Depending on which aspects of the orbit we know, we can use these equations to ultimately solve for true anomaly.

Before we wrap up this chapter, we wanted to highlight that all of these orbital elements a, e, i, Ω - omega, ω - omega, ν - nu sound kind of like the English vowels a, e, i, o (for both omegas!), and u.

We are also leaving you with this handy-dandy table to help you remember the symbolic representations of all these orbital elements!

Table 2.2: Keplerian Orbital Element Symbols

Symbol	Definition
a	Semi-Major Axis
e	Eccentricity
i	Inclination
Ω	RAAN (Right Ascension of Ascending Node)
ω	Argument of Perigee
ν	True Anomaly

Chapter 3

Orbital Mechanics: 2D

In this chapter, we will explore Kepler's and Newton's Laws, which broadly describe the motion of orbiting bodies. Johannes Kepler and Sir Isaac Newton, illustrated in Figure 3.1, were mathematicians whose scientific work formed the fundamentals of orbital mechanics. Let's start in the 17th century with Kepler!

Figure 3.1: Kepler and Newton

3.1 Kepler's Laws of Planetary Motion

German astronomer, mathematician, and philosopher Johannes Kepler published his laws of planetary motion in the early 1600s [17]. These laws came about when a transition from geocentrism, the belief that everything in the universe rotates around the Earth, to heliocentrism, the belief that everything rotates around the Sun [17] was taking place. Today, these laws are still used to help us understand the mechanics of our universe. We'll use Figure 3.2 to help navigate the meaning of these three laws in how they describe planetary motion around the Sun.

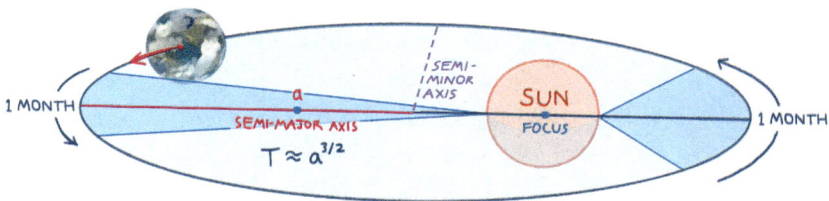

Figure 3.2: Kepler's Laws

Kepler's Laws of Planetary Motion [14]:

- **Kepler's First Law:** The planets move in a plane; the orbit of each planet is an ellipse with the Sun at one focus.

- **Kepler's Second Law:** The **law of areas** dictates that a satellite will cover a constant area over a given amount of time. A result of this law is that objects closer to the Sun (focus) travel faster than those farther away from the Sun.

- **Kepler's Third Law:** The ratio of the square of the orbital period, T (other sources may refer to T as P), of a planet around the Sun to the cube of the semi-major axis, a, of the ellipse is the same for all planets: $T \approx a^{\frac{3}{2}}$.

Kepler's laws elegantly describe the motion of the planets, but they offer no insight as to why the planets move in this way. Newton extended the work of Kepler, particularly Kepler's Third Law, with his Laws of Motion and the **universal law of gravitation** [23].

Doggo Help

Woof! Kepler's First Law explains that Earth travels in an elliptical orbit around the Sun. Remember, we use the following parameters when thinking about elliptical orbits:

- The dimension of an ellipse is described by its major and minor axis. With orbits, we use semi-major axis, a, and eccentricity, e.

- The semi-major axis, a, is the average of the radius at perigee and apogee. For circular orbits, the radius to apogee and perigee are equal! Therefore, a is simply the radius of the orbit, where the radius is $r = h + R_e$, with h as the altitude and R_e as the Earth's radius. Remember to add in the Earth's radius! Many often forget this critical step.

3.2   Newton's Law of Gravitation

Newton's Law of Gravitation states that two bodies of mass, m and M, attract each other with a force that is proportional to their masses and that is inversely proportional to the square of the distance r between them.

$$F = \frac{GMm}{r^2} \text{ [N]} \tag{3.1}$$

In the above equation, G is the universal constant of gravitation, equal to $6.67 \cdot 10^{-11}$ Nm^2/kg^2 and M_E, or the mass of the Earth, is $5.974 \cdot 10^{24}$ kg. Earth's gravitational parameter, μ (pronounced mu), is equal to 398600 km^3/s^2, and is found by multiplying the universal constant of gravitation, G, by Earth's mass. In other words, we define $\mu = GM_E$.

Consider a satellite orbiting Earth. Since Earth's gravitational force is inversely proportional to the distance between the Earth and this satellite, Earth's gravity pulls more on the satellite when it is closer to the Earth. This is why the satellite travels fastest at perigee!

3.3 Orbital Period

The time required for a satellite to complete an orbit is known as the **orbital period**, *T*. For a circular orbit, speed is constant, so the period is equal to the circumference, mapped out by the orbit, divided by the speed. For elliptical orbit, the equation takes a similar, but slightly different, form:

$$T = \frac{2\pi}{\sqrt{\mu}} a^{\frac{3}{2}} \ [\text{sec}] \qquad (3.2)$$

Provided that μ and a use the same length unit like kilometers, Equation 3.2 returns the orbital period in seconds. As we previously mentioned, for Earth-orbiting satellites, it is often most useful to convert this period into hours or minutes.

The equation for orbital period relates back to Kepler's Laws. Remember Kepler's Third Law explains that the time period, *T*, of a planet revolving around the Sun equals the square root of the cube of the semimajor axis. Let's take a moment with Sat and try to compute the orbital period for a few different orbits. Flip back a few pages if you don't remember Kepler's third law, and see if you can identify the similarity between the law and the equation for orbital period.

Example with Sat!

Compute the orbital period for the following satellite systems:

1. GEO satellite (altitude = 35,786 km)

2. SES O3b satellite in MEO (altitude = 8,062 km)

3. Telesat Lightspeed satellite in LEO (altitude = 1,000 km)

Solution:

To compute the orbital period of each satellite, we need the semi-major axis, a, which we can then plug into the orbital period equation shown in Equation 3.2.

1. The altitude of a GEO satellite is 35,786 km, so its semi-major axis, a, is 35,786 km + 6,378 km = 42,164 km. Since eccentricity, e, is zero for GEO satellites and $r_a = r_p$, we obtain:

$$T = \frac{2\pi}{\sqrt{\mu}} a^{\frac{3}{2}} = \frac{2\pi}{\sqrt{398600 \text{ km}^3/\text{s}^2}} (42164 \text{ km})^{3/2}$$

$$\approx 86164 \text{ s} \approx 23.9 \text{ hours}$$

2. Given that an SES O3b satellite orbits at an altitude of 8,062 km, we must add the earth's radius, 6,378 km, to find the semi-major axis to be 14,440 km. Using the same orbital period equation, we compute:

$$T = \frac{2\pi}{\sqrt{\mu}} a^{\frac{3}{2}} = \frac{2\pi}{\sqrt{398600 \text{ km}^3/\text{s}^2}} (14440 \text{ km})^{3/2}$$

$$\approx 17269 \text{ s} \approx 4.8 \text{ hours}$$

Example with Sat!

3. A Telesat Lightspeed satellite in LEO orbits at an altitude of 1,000 km. Accounting for the earth's radius yields a semi-major axis of 7,378 km. Plugging this semi-major axis into the orbital period equation gives us:

$$T = \frac{2\pi}{\sqrt{\mu}}a^{\frac{3}{2}} = \frac{2\pi}{\sqrt{398600 \text{ km}^3/\text{s}^2}}(7378 \text{ km})^{3/2}$$

$$\approx 6307 \text{ s} \approx 1.75 \text{ hours}$$

3.4 The Two-Body Problem

In orbital mechanics, the classic two-body problem refers to solving the equations of motion of two separate bodies due solely to their mutual gravitational attraction [4].

The Laws of Conservation constrain the two-body equations of motion. Specific angular momentum, h, or the angular momentum of a body divided by its mass, is conserved: $\frac{dh}{dt} = 0$, thus the body's motion becomes confined to a plane. Specific angular momentum is measured in units of $\frac{\text{m}^2}{\text{kg} \cdot \text{s}}$ and is important because it remains constant over an orbit in ideal conditions. Note that when we say "specific," we mean per kg. To find specific angular momentum and solve the two-body problem, we use the following equation:

$$\vec{h} = \frac{\vec{L}}{m} = \vec{r} \times \vec{v} \ [\text{m}^2/\text{s}] \tag{3.3}$$

Specific angular momentum is also found by taking the cross product of the relative position vector, $\vec{r}$, and the relative velocity vector, $\vec{v}$.

Doggo Help

Woof! Looking at Equation 3.3, we can see that the specific angular momentum $\vec{h}$ is directly related to $\vec{r}$, the radius of the orbit, and the velocity of the orbiting object, $\vec{v}$. This means that as the radius decreases, velocity will increase to keep the specific angular momentum constant. This is why satellites travel fastest at perigee!

According to the Law of Conservation of Energy, energy is conserved in an isolated system. In the context of a satellite orbiting Earth, energy is neither gained nor lost, only exchanged between kinetic and potential forms, and the sum total is constant. This is represented by the specific total energy, ϵ:

$$\text{Energy is constant at any point: } \epsilon = -\frac{\mu}{2a} \text{ [J/kg]} \qquad (3.4)$$

The specific total energy plays an important role in defining the Vis-viva ("living force") equation:

$$\text{Vis-viva equation: } \epsilon = \frac{v^2}{2} - \frac{\mu}{r} \text{ [J/kg]} \qquad (3.5)$$

where the radius, r, is equal to the sum of the radius of the Earth, 6378 km, plus the orbital altitude and that a is the semi-major axis. The Vis-viva equation in Equation 3.5 is derived from the sum of kinetic and potential energy.

$$E_{\text{Total}} = E_{\text{Kinetic}} + E_{\text{Potential}} \qquad (3.6)$$

Recall from mechanics that the equation for the kinetic energy of an object is $\frac{1}{2}mv^2$, where m is the object's mass and v is the object's velocity. Additionally, its gravitational potential energy equals mgh, where g is the acceleration of gravity on Earth's surface and h is the *altitude* of that object or its distance from the center of the Earth. If we substitute E_{Kinetic} and $E_{\text{Potential}}$ with these formulas, we obtain:

$$E_{\text{Total}} = \frac{1}{2}mv^2 + mgh \text{ , where } g = -\frac{GM}{r^2} \qquad (3.7)$$

Substituting g yields:

$$E_{\text{Total}} = \frac{1}{2}mv^2 + m\left(-\frac{GM}{r^2}\right)r \qquad (3.8)$$

We can rewrite the second component as $-\frac{mGMr}{r^2}$. Notice how there is an r in the numerator and r^2 in the denominator. This means we can simplify by eliminating an r term from both the numerator and the denominator! This simplification gives us the following result:

$$E_{\text{Total}} = \frac{1}{2}mv^2 - \frac{mGM}{r} \tag{3.9}$$

Now we have an equation for the total energy of any object of mass, m, orbiting our Earth! However, we want to solve for the specific total energy, ϵ, which means we must account for mass by dividing the energy of our object, E_{Total}, by its mass, m.

$$\epsilon = \frac{E_{\text{Total}}}{m} = \frac{v^2}{2} - \frac{GM}{r} \tag{3.10}$$

Notice how we took the equation from Equation 3.9 and canceled out an m from each term, or in essence, divided each term by m. From here, since we know that $\mu = GM$, we obtain the Vis-viva equation from Equation 3.5.

$$\epsilon = \frac{v^2}{2} - \frac{\mu}{r} \tag{3.11}$$

Next, we will set the specific total energy in Equation 3.4 equal to the Vis-viva equation in Equation 3.5 and solve for velocity to derive the orbital velocity equation.

$$-\frac{\mu}{2a} = \frac{v^2}{2} - \frac{\mu}{r} \tag{3.12}$$

A particular type of orbital velocity is the escape velocity, which is the minimum velocity required for a free, non-propelled object to escape from the gravitational influence of a massive body. To compute the escape velocity, set the equation for conservation of energy equal to the Vis-viva equation and take the limit as $a = \infty$. Note that when we take the limit as $a = \infty$, we set the energy side of the equation to 0 (because dividing by a huge number produces a small number).

$$0 = \frac{v^2}{2} - \frac{\mu}{r} \tag{3.13}$$

Now, to solve for escape velocity, all we need to do is solve for v. Adding $\frac{\mu}{r}$ to both sides of the equation:

$$\frac{v^2}{2} = \frac{\mu}{r} \tag{3.14}$$

We obtain the escape velocity if we multiply both sides by 2 and take the square root to eliminate the exponent.

$$v_{escape} = \sqrt{\frac{2\mu}{r}} \tag{3.15}$$

History with Squit!

The Soviet satellite Luna 1 was the first satellite to escape Earth's orbit in 1959. However, this was actually an accident! The lunar probe was meant to crash into the moon, but it missed and instead became the first man-made satellite to orbit the Sun [30]!

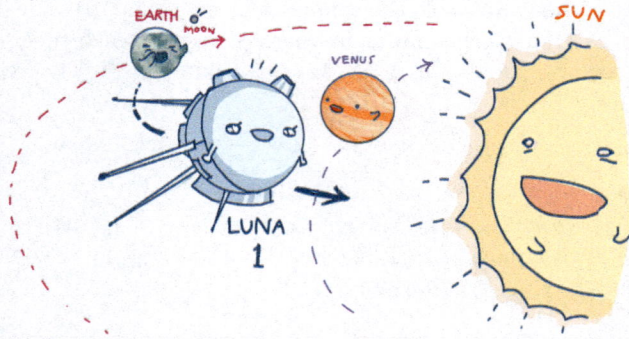

Example with Sat!

Find the escape velocity for an object:

1. On the surface of the Earth

2. Orbiting at an altitude of 500 km

Solution:

1. If an object is on the surface of the Earth, then the altitude is equal to 0, so $r = 6378$ km. Let's set the energy equation equal to the Vis-viva equation (defined in Equation 3.5) and solve for v.

$$-\frac{\mu}{2a} = \frac{v^2}{2} - \frac{\mu}{r}$$

$$0 = \frac{v^2}{2} - \frac{\mu}{r}$$

$$\frac{v^2}{2} = \frac{\mu}{r}$$

$$v = \sqrt{\frac{2 \times 398600 \text{ km}^3/\text{s}^2}{6378 \text{ km}}}$$

$$v_{escape} \approx 11.2 \text{ km/s}$$

2. If an object is orbiting at 500 km, then $r = 6378$ km $+$ 500 km $= 6878$ km, and the escape velocity becomes:

$$\frac{v^2}{2} = \frac{\mu}{r}$$

$$v = \sqrt{\frac{2 \times 398600 \text{ km}^3/\text{s}^2}{6878 \text{ km}}}$$

$$v_{escape} \approx 10.8 \text{ km/s}$$

3.5 The Orbit Equation

The **orbit equation** defines the path of the satellite of mass, m, around the Earth of mass, M_E without having to specify a satellite's position as a function of time.

$$r(\theta) = \frac{h^2}{\mu} \frac{1}{(1 + e\cos\theta)} \text{ [km]} \qquad (3.16)$$

The specific angular momentum, h, of the satellite, the Earth's gravitational parameter, μ, and eccentricity, e, of the orbit are constant, and the true anomaly, θ, also referred to as ν, is the angle between perigee and the position of the satellite. We can also write the Orbit Equation in terms of semi-major axis:

$$r(\theta) = a\frac{1 - e^2}{(1 + e\cos\theta)} \text{ [km]} \qquad (3.17)$$

When we set θ equal to 0, we compute the radius of perigee. Similarly, when we set $\theta = 180$ degrees, we compute the radius of apogee.

Example with Sat!

Use the orbit equation in Equation 3.17 to compute the radius of a Telesat Lightspeed satellite at a true anomaly of $\theta = 72°$.

Solution:
Plug in $\theta = 72°$ into the orbit equation, along with an eccentricity $e = 0$ (since we know that $r_a = r_p$), and a semi-major axis of $a = 1000 \text{ km} + 6378 \text{ km} = 7378 \text{ km}$.

$$r(\theta) = a\frac{1 - e^2}{(1 + e\cos\theta)}$$

$$= \frac{1 - 0^2}{(1 + 0\cos(72°))}$$

$$= 7378 \text{ km}$$

3.6   Ground Tracks

A ground track is the projection of a satellite's orbit onto the surface of the Earth. At a point in time, imagine a line from the center of the Earth to the satellite. Where the line intersects the Earth's surface is a point on the ground track.

The ground track of a LEO satellite oscillates around the Earth's equator every orbit, reaching a maximum and minimum latitude. On a Mercator projection, the ground track resembles a sine wave. If the Earth did not rotate, the ground track would resemble a sinusoidal-like track, repeatedly traced as the satellite orbited the Earth. The Earth rotates eastward beneath the satellite orbit at $15.04°$per hour, so the ground track advances westward at the same rate. The distance between two consecutive crossings of the equator is the amount the Earth rotates in one orbit of the satellite. Therefore, from the ground track, we can compute the period of the satellite.

Figure 3.3 depicts the ground tracks of LEO, GEO, and geosynchronous orbits.

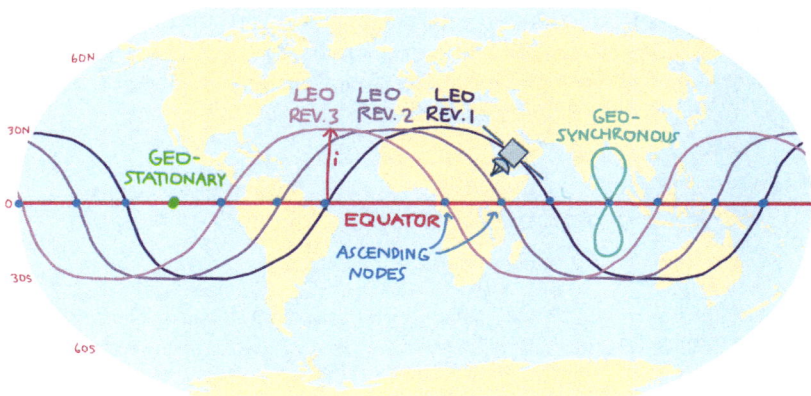

Figure 3.3: Ground Tracks

We can also use ground tracks to infer other orbital parameters such as inclination. If a satellite has an inclination of zero, its ground track will be a flat line over the equator. Otherwise, the inclination can be calculated from the latitudinal distance between the equator and the maximum or minimum of the ground track's sinusoidal pattern.

> **Doggo Help**
>
> *Bark!* An easy way to remember which way Earth rotates is
> that the Sun rises in the east and sets in the west. Therefore,
> the Earth must be moving from west to east as the east sees
> the sunrise before the west!

3.7   Bonus Material: Two-Line Elements

A two-line element (TLE) set is a standard data format specified by
North American Aerospace Defense Command (NORAD) that contains
a list or orbital elements of an Earth-orbiting object for a given point
in time, or *epoch*. TLEs are used to predict the precise state, both the
position and the velocity, of a satellite at any point in the past or future.
One critical TLE use case is to predict the future orbital track of space
debris in order to conduct risk analysis, close approach analysis, and
collision avoidance maneuvers. The United States Space Force (USSF)
tracks objects in Earth's orbit and creates a corresponding TLE for each
object. TLEs are publicly available on the Space Track and Celestrak
websites, and consist of a title line followed by two lines of formatted
text. Consider the following example TLE:

```
ISS (ZARYA)

1 25544U 98067A 08264.51782528 -.00002182 00000-0
-11606-4 0 2927

2 25544 51.6416 247.4627 0006703 130.5360 325.0288
15.72125391563537
```

The title line with the satellite's name is optional since each data line
includes a unique object identifier. The characteristic data included in
the next two lines describes the following:

First Line of TLE Formatted Text

- **01**: Line number: 1
- **03-07**: Satellite number: 25544
- **08**: Classification (U = Unclassified, C = Classified, S = Secret)
- **10-17**: International Designator (international identifier assigned to every object in space, which includes the year it was launched (98 means 1998), a three-digit incrementing launch number for that year (067 meaning this satellite was the 67th successful launch in 1998), and a three character code representing which part of the sequence was launched for that object (A means that this was the *first* launch for this satellite. Some satellites or bodies in space may require multiple launches for different pieces of the same body).
- **19-32**: Epoch year and day (last two digits of the year and the day of the year including a fractional portion of day): 08264.5178258
- **34-43**: First derivative of the mean motion (angular speed of the satellite): -.00002182
- **45-52**: Second derivative of the mean motion (angular velocity of the satellite): 00000-0
- **54-61**: Drag term of BSTAR, which is a way of modeling aerodynamic drag on a satellite: -11606-4
- **63**: Satellite's ephemeris number (ephemeris provides the trajectory of orbiting bodies in space): 0
- **65-69**: Element number (counter of how many times a TLE has been generated for this satellite) and checksum (generated to account for an error in the message):   2927

Second Line of TLE Formatted Text

- **01**: Line number: 2
- **03-07**: Satellite number: 25544
- **09-16**: Inclination (degrees): 51.6416
- **18-25**: RAAN (degrees): 247.4627
- **27-33**: Eccentricity: 0006703
- **35-42**: Argument of perigee (degrees): 130.5360

- **44-51**: Mean anomaly (degrees): 325.0288
- **53-63**: Mean motion (revolutions per day): 15.72125391
- **64-68**: Number of revolutions at current epoch (how many times it has revolved since launch): 56353
- **69**: Checksum: 7

Note that TLEs are NOT meant for parabolic or hyperbolic orbits due to format limitations. If you'd like to learn more about TLEs, the Space Track website offers a "TLE Format" section at space-track.org, providing an overview of each portion of the TLE.

3.8 Problem Set 2: Orbital Mechanics 2D

Problem	Topic	Points
1	Earth's Gravitational Parameter	3
2	Orbital Speed Sketch	3
3	Vis-Viva	3
4	Orbital Velocity Computation	3
5	Orbital Period Computation	3
6	Orbital Period Concept	3
Total:		18

Exercise 3.1

(**3 points**) Perform unit analysis (write the product of the units and show cancellation) of Earth's gravitational parameter, $\mu = GM = 3.986 \cdot 10^{14} \ \mathrm{m}^3/\mathrm{s}^2$. Note, however, that we often see $\mu = 398,600 \mathrm{km}^3/\mathrm{s}^2$, so your final answer should be in these units!

Exercise 3.2

(**3 points**) At what point in an elliptical orbit is the satellite's speed the greatest? Where on the orbit is the satellite's speed the least? Include a sketch and label the relevant orbital parameters.

Exercise 3.3

(**3 points**) Given the Vis-viva equation and the fact that energy is constant at any point, solve for the velocity of the orbit at any point. Use this equation to compute the velocity of a satellite in a circular orbit at 400 km.

Exercise 3.4

(**3 points**) Compute the minimum and maximum velocity for a:

1. MEO satellite with a perigee and apogee of 8,200 km.

2. LEO satellite with a perigee of 500 km and an apogee of 4,000 km.

Exercise 3.5

(**3 points**) Use the orbit equation to compute the orbital radius of a MEO satellite with a true anomaly of 60°, a perigee of 27,355 km, and an apogee of 44,221 km.

Exercise 3.6

(**3 points**) Assuming your audience has no orbital mechanics background, write an explanation for the concept of orbital period, where its equation comes from, and why the value of the orbital period for a geostationary orbit is meaningful. Then explain why a communications satellite might utilize a geostationary orbit?

Chapter 4

Hohmann Transfers

A Hohmann transfer is the most fuel-efficient, two-burn maneuver to move a satellite from one circular orbit to another. However, it is not necessarily the fastest [8].

In a Hohmann transfer, the first burn maneuvers the spacecraft into the elliptical transfer orbit (shown in pink in Figure 4.1), meaning that it boosts the apogee of the spacecraft, via an elliptical transfer orbit, to that of the desired second circular orbit, this transfer orbit can alternatively be thought of as "pushing" or "pulling" one of the foci away from the Earth, converting a circular orbit into an elliptical orbit. The second burn raises the perigee of the elliptical transfer orbit to that of the final desired circular orbit.

To simplify the problem, we assume that these burns are *impulsive*, which is a brief firing of the on-board propulsion unit of a spacecraft that instantaneously causes a change in its velocity (Δv) in either speed, direction, or both [27]. This assumption is primarily valid for high-thrust systems with short burn times compared to the vehicle's coasting time.

To accelerate in the forward direction requires the thruster to eject fuel (and therefore impose a force) opposite to the direction of motion! This burn also occurs *in the direction of flight*, meaning that the spacecraft is accelerating in the direction it is currently flying, tangential to the transfer orbit the spacecraft enters or exits. The elliptical transfer orbit must therefore intersect or be tangent to both of the two circular orbits [3].

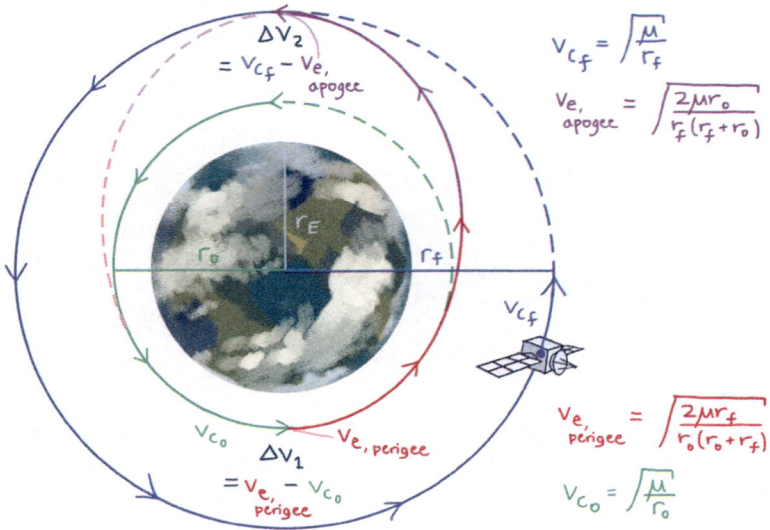

Figure 4.1: Hohmann Transfer

The change in velocity, Δv_1, required for the first burn is the difference between the velocity at perigee of the elliptical transfer orbit and the velocity of the initial circular orbit. Some people use the phrase "final minus initial orbit" to help them remember the order of the two velocity terms.

$$\Delta v_1 = v_{\text{elliptical}_{\text{perigee}}} - v_{\text{circular}_{\text{initial}}} \tag{4.1}$$

As we have previously seen, to derive the equation for velocity at a point in an orbit, we set the vis-viva and energy equations, which are represented by $\epsilon = \frac{v^2}{2} - \frac{\mu}{r}$ and $\epsilon = -\frac{\mu}{2a}$, respectively, equal to each other, and then isolate v:

$$\frac{v^2}{2} - \frac{\mu}{r} = -\frac{\mu}{2a} \tag{4.2}$$

$$\frac{v^2}{2} = \frac{\mu}{r} - \frac{\mu}{2a} \tag{4.3}$$

$$v^2 = \frac{2\mu}{r} - \frac{2\mu}{2a} \tag{4.4}$$

Ultimately, we find that the velocity of a spacecraft at an orbital radius, r, is equal to:

$$v = \sqrt{\mu(\frac{2}{r} - \frac{1}{a})} \tag{4.5}$$

History with Squit!

Walter Hohmann was born on March 18, 1880 in a small town in Western Germany. He was actually a civil engineer by education and lectured about statics and bridge building, but he had a passion for space [28]. He authored a book in 1925 titled "The Attainability of Heavenly Bodies" [15]. This book tackled five major barriers to spaceflight: leaving Earth, returning to Earth, flight in free space, circumnavigating celestial bodies, and landing on celestial bodies. Hohmann demonstrates that you can contribute to this field from any background!

Equation 4.5 will be used to calculate the velocities for both our elliptical orbits and circular orbits involved in the Hohmann transfer. In the case of the first burn of a Hohmann Transfer, the radius of perigee is equal to the radius of our initial circular orbit, r_0, and the radius of apogee is equal to the radius of our final circular orbit, r_f. To compute the velocity at the location of our "initial orbit", which in this case is the perigee of the elliptical transfer orbit, we plug in $r = r_0$ in the first term. For the second term, we substitute r_0 and r_f into our semimajor axis equation, $a = \frac{r_0 + r_f}{2}$. Using the semimajor axis enables us to bring

the equation for velocity in terms of r and simplify. Plugging these two values into the first and second terms yields the following:

$$v_{elliptical_{perigee}} = \sqrt{\frac{2\mu}{r_0} - \frac{2\mu}{r_0 + r_f}} \tag{4.6}$$

Given there is a common numerator of 2μ in both terms in the expression. Let's factor it out.

$$v_{elliptical_{perigee}} = \sqrt{2\mu \left(\frac{1}{r_0} - \frac{1}{r_0 + r_f}\right)} \tag{4.7}$$

Let's work to find a common denominator by multiplying the numerator and denominator by $(r_0 + r_f)$ in the first term and (r_0) in the second term:

$$v_{elliptical_{perigee}} = \sqrt{2\mu \left(\frac{(r_0 + r_f)}{r_0(r_0 + r_f)} - \frac{(r_0)}{(r_0 + r_f)(r_0)}\right)} \tag{4.8}$$

Now, since we have a common denominator, we can subtract these fractions and simplify.

$$v_{elliptical_{perigee}} = \sqrt{2\mu \left(\frac{r_0 + r_f - r_0}{r_0(r_0 + r_f)}\right)} \tag{4.9}$$

Finally, simplify the numerator ($r_0 - r_0 = 0$) and multiply the 2μ back into the expression.

$$v_{elliptical_{perigee}} = \sqrt{\frac{2\mu(r_f)}{r_0(r_0 + r_f)}} \tag{4.10}$$

Great job! We have the equation for velocity of our first burn, and we know that the first burn of a Hohmann transfer maneuvers the spacecraft from the initial circular orbit to the elliptical transfer orbit. As a result, to compute the Δv we must find the velocity of the spacecraft in the initial circular orbit.

Doggo Help

But wait! You may be wondering why did our initial burn create an elliptical orbit with the same perigee and a raised apogee instead of a larger circular orbit. We have an elliptical orbit because we added velocity to our spacecraft with the first burn, but not enough energy to escape Earth's gravity to a higher altitude! Therefore, our satellite will swing out further in an elliptical orbit because of the added energy, but it will need a second burn to properly escape Earth's gravity into a circular orbit.

To compute the velocity of the initial circular orbit, we know that $a = r = r_0$; therefore, when we plug this into the second term of our orbital velocity equation, we have a common denominator of r_0, and the following holds true:

$$v_{\text{circular}_{\text{initial}}} = \sqrt{\frac{2\mu}{r_0} - \frac{\mu}{r_0}} \tag{4.11}$$

$$v_{\text{circular}_{\text{initial}}} = \sqrt{\frac{\mu}{r_0}} \tag{4.12}$$

To compute Δv_1, we substitute the equation for the velocity at perigee of our elliptical transfer orbit and the velocity of our initial circular orbit:

$$\Delta v_1 = v_{\text{elliptical}_{\text{perigee}}} - v_{\text{circular}_{\text{initial}}} \tag{4.13}$$

$$\Delta v_1 = \sqrt{\frac{2\mu r_f}{r_0(r_0 + r_f)}} - \sqrt{\frac{\mu}{r_0}} \ [\text{m/s}] \tag{4.14}$$

where r_0 represents the radius of our initial circular orbit, r_f represents the radius of our final elliptical transfer orbit, and μ is the celestial body's gravitational parameter.

It is important to note that we must still include the radius of the Earth, 6378 km, when calculating r_0 and r_f and that μ must share the same units of distance as r_0 and r_f (i.e. if r_0 and r_f are in kilometers, then μ must be calculated with units of km^3/s^2, *not* m^3/s^2)!

A second burn is required to convert the elliptical transfer orbit into the final circular orbit. As previously stated, this burn raises the *perigee* of the elliptical transfer orbit to that of the final circular orbit. As before, we assume this burn is impulsive and occurs in the direction of motion. If this burn were not performed, our spacecraft would remain in the elliptical transfer orbit. The velocity change necessary for the second burn, Δv_2, is the difference in the velocity of the outer, final circular orbit and the velocity at *apogee* of the transfer orbit.

$$\Delta v_2 = v_{\text{circular}_{\text{final}}} - v_{\text{elliptical}_{\text{apogee}}} \tag{4.15}$$

$$\Delta v_2 = \sqrt{\frac{\mu}{r_f}} - \sqrt{\frac{2\mu r_0}{r_f(r_f + r_0)}} \ [\text{km}] \tag{4.16}$$

If we wanted to perform a Hohmann transfer to a lower circular orbit, we would follow these same steps, but instead *retrofire* our engines, or burn in the opposite direction of flight.

Example with Sat!

What is the required Δv_1 and Δv_2 for a Hohmann Transfer from 500 km to an outer circular orbit of 1200 km? Can you find Δv if this is the only maneuver you need to perform?

Solution:
Using Equations 4.14 and 4.16, we can simplify Δv_1 and Δv_2 based on what we know about r_0 and r_f, which are $r_0 = 500$ km + 6378 km = 6878 km and $r_f = 1200$ km + 6378 km = 7578 km, respectively. Note that r_0 and r_f have units of km and $\mu = 398600$ km^3/s^2:

$$\Delta v_1 = \sqrt{\frac{2\mu r_f}{r_0(r_0 + r_f)}} - \sqrt{\frac{\mu}{r_0}}$$

$$= \sqrt{\frac{2(398600 \text{ km}^3/\text{s}^2)(7578 \text{ km})}{6878 \text{ km}(6878 \text{ km} + 7578 \text{ km})}} - \sqrt{\frac{398600}{6878 \text{ km}}}$$

$$= 0.182 \text{ km/s}$$

$$\Delta v_2 = \sqrt{\frac{\mu}{r_f}} - \sqrt{\frac{2\mu r_0}{r_f(r_f + r_0)}}$$

$$= \sqrt{\frac{398600}{7578 \text{ km}}} - \sqrt{\frac{2(398600 \text{ km}^3/\text{s}^2)6878 \text{ km}}{7578 \text{ km}(7578 \text{ km} + 6878 \text{ km})}}$$

$$= 0.178 \text{ km/s}$$

To find Δv, we can sum Δv_1 and Δv_2:

$$v = \Delta v_1 + \Delta v_2$$
$$= 0.182 \text{ km/s} + 0.178 \text{ km/s}$$
$$= 0.36 \text{ km/s}$$

4.1   Problem Set 3: Hohmann Transfers

Problem	Topic	Points
1	Delta-v Derivation	3
2	Hohmann Calculator	3
3	Hohmann Transfer from LEO to GEO	3
4	Hohmann and Mean Anomaly Calculator	3
5	Interplanetary Hohmann Transfer	3
Total:		15

Exercise 4.1

(**3 points**) In this chapter, we derived equations for Δv_1 for a Hohmann Transfer. Use a similar process to derive Δv_2.

Exercise 4.2

(**3 points**) In Matlab, write a script that computes the Δv's of each Hohmann Transfer burn and the total delta-v of the maneuver. Make sure your Hohmann Transfer Calculator code computes the two individual Δv's as well as the total Δv of the transfer, that is the sum of Δv_1 and Δv_2. If you do not have Matlab, consider implementing these computations in a program like Excel.

Compare your code with a partner. When you do this, focus on the way in which you communicate information to your partner and also focus on listening to your partner. Share the structure of your code. What are the inputs to your calculator? What are your outputs? What are their inputs and outputs?

Compare the results of your code for a maneuver from an:

1. Inner circular orbit of 1000 km to an outer circular orbit of 3000 km

2. Outer circular orbit of 3000 km to an inner circular orbit of 1000 km

How do the results of your calculator's output compare to those of your partner's?

Did you implement any improvements after seeing your partner's code?

Exercise 4.3

(**3 points**) Using your Hohmann Transfer Calculator, compute the total Δv required for a transfer orbit from a 400 km LEO parking orbit to GEO. Quantify and output the semi-major axis and period of the three orbits. What do you notice about the period of the geostationary orbit?

Exercise 4.4

(**3 points**) Upgrade your Hohmann Transfer calculator to also include Mean Anomaly computations and propagate your maneuvers in time (completing a full rotation in your first orbit, your transfer, and a full rotation in your second orbit). Your code should output the delta-v's and the total time for this maneuver to occur.

Exercise 4.5

(**3 points**) Let's consider the Hohmann transfer for an interplanetary maneuver between Earth and Mars, using Earth's orbit around the Sun as the initial orbit and Mars' orbit around the Sun as the final orbit.

The distance between the Sun and the Earth is defined as

one astronomical unit (AU), or 147 million km. The distance between the Sun and Mars is 233 million km. The gravitational parameter of the Sun, μ_{Sun}, is $1.327 * 10^{11}$ km^3/s^2, and the radius of the Sun is 696,340 km.

Compute the period of both the inner and outer circular orbits, as well as the time it takes for the transfer between the two. What is the Δv_1 and Δv_2 for this Hohmann transfer?

How do we know quantitatively when to conduct our burn so we don't miss Mars? Note: You may need to conduct further research to answer the last question. We recommend investigating the timing of interplanetary Hohmann transfers between Earth and Mars.

Chapter 5

Space Sustainability

As of May 2021, the United States Space Force had cataloged 22,485 space debris objects larger than approximately 10 cm in diameter, 13,176 of which were in Low Earth Orbit (LEO)[12]. In 2019, the European Space Agency (ESA) estimated that there were likely around 130 million total objects larger than 1 mm [32]. This is why it's always important to pick up after yourself!

Figure 5.1: American and European Debris Tracking Efforts

5.1 Orbital Debris Mitigation

Since the early 1960s, more than 14,000 objects have been launched into space [12]. The year 2022 marked a historic peak with 2,163 objects launched globally, 83.0% of which were U.S. owned [32]. This increase in launches has led to a rise in the amount of orbital debris. Logically, as more objects occupy space, the risk of collision and creation of additional debris also increases. Due to the high velocity of objects in Earth's orbit, often many times faster than the speed of a bullet, even debris smaller than a centimeter can cause serious damage to spacecraft and other satellites, potentially rendering critical systems nonfunctional [9].

Coined in 1978 by Dr. Don Kessler, the Kessler Syndrome defines a worst-case scenario in which the critical mass of in-orbit objects catalyzes a chain reaction of collisions that, in turn, create more debris and leads to follow-on collisions in a snowball effect [18].

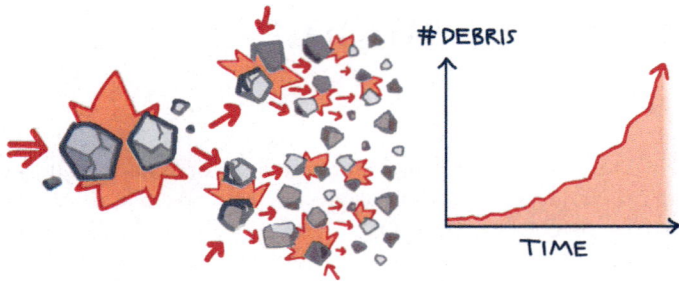

Figure 5.2: Kessler Syndrome

Proper post mission disposal (PMD) operation and regulation are essential to prevent the accumulation of space debris, as failure to address this issue could make it nearly impossible to launch and restore satellite networks. The three main agencies currently involved in regulating PMD within the U.S. are the Federal Communications Commission (FCC), the National Aeronautics and Space Administration (NASA), and Congress. To address the issue of space debris, numerous tracking and mitigation initiatives have been established, including the revision of the United States' Orbital Debris Mitigation Standard Practices (ODMSP) [21].

In 2022, the FCC shortened the time frame for deorbiting LEO satellites to within five years of mission completion, from the prior 25-year deorbit limit [10]. Satellites at altitudes below approximately 550 km and satellites that are equipped with certain propulsive or drag technologies can generally comply with this regulation. In contrast, satellites at higher altitudes require active plans involving fuel storage and a final end-of-mission maneuver, or the potential use of debris removal services. Additionally, under part 25 of the Code of Federal Regulations (CFR), the Commission requires operators to ensure that their satellites burn up completely during re-entry to avoid any risk of human casualty.

5.2 Atmospheric Re-entry

Even though satellites largely exist in a vacuum, friction-like mechanisms cause **orbital decay**, which gradually saps the kinetic energy of a satellite's orbital motion by dissipating it as mechanical, gravitational, or electromagnetic energy. The most prevalent of such force on LEO spacecraft is **atmospheric drag**, which is a friction force directly proportional to the density of the atmospheric. As a result, the primary method of satellite disposal in LEO and in some portions of MEO is atmospheric re-entry.

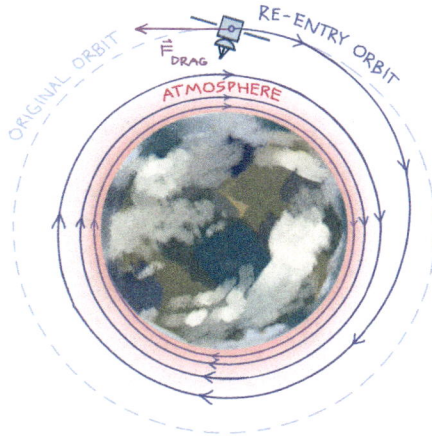

Figure 5.3: Atmospheric Drag

Solar activity can also impact the rate of a satellite's orbital decay. During the eleven-year solar cycle, the Sun's activity at solar maximum, a multi-year period of increased magnetic activity leading to more solar radiation and winds, causes greater drag compared to solar minimum [13].

Atmospheric re-entry can be performed actively or passively. When performed actively, satellites maneuver to lower their altitude to take advantage of drag and deorbit more quickly. If an operator passively deorbits their system, they plan for natural orbital decay. Ultimately, these spacecraft burn up in Earth's atmosphere instead of becoming space hazards or moving to a graveyard orbit. Satellites launched to or around the ISS altitude ($\sim$400 km) naturally decay in under five years. At orbits beyond 600 km, decay within five years can no longer be guaranteed [31].

5.3   Graveyard Orbits

Maneuvering the satellite to a graveyard orbit is another type of PMD. A graveyard orbit is an orbit that lies outside of typical operational orbits. The most common is slightly above GEO, where many ghost satellites, which are no longer operational, are positioned.

History with Squit!

A ghost satellite doesn't mean a spooky satellite that has risen from the dead! The term refers to a satellite that is no longer operational due to mission completion, regulatory changes, or technical malfunction.

For GEO satellites, the graveyard orbit is located approximately 300 km above GEO altitude. To maneuver a GEO satellite to this orbit, a change in velocity of approximately 11 m/s, or three months' worth of fuel for regular station keeping is needed [24]. This implies that satellite operators must end their missions before they run out of fuel, reducing the time available to generate revenue from providing services to their customers.

5.4   In-orbit Servicing

If companies aren't proactive and defer moving their satellites to the graveyard orbit, they run the risk of not having enough fuel to maneuver the satellite out of the way of other satellites in GEO. Moreover, if companies use a service to refuel and extend the life of their satellite beyond its original planned lifespan, the likelihood of a different failure occurring during this prolonged lifespan increases. If a satellite malfunctions before it can move to the graveyard orbit, an in-orbit servicing system may be utilized to maneuver the satellite out of GEO.

In-orbit servicing offers an alternative way to complete PMD by re-trieving a particular spacecraft and re-positioning it to either a grave-yard orbit or a LEO orbit for atmospheric re-entry. Satellite companies must consider the trade of the extra revenue they might generate versus the cost of PMD when deciding whether to extend a satellite's mission lifetime.

Doggo Help

Congrats!!!! Doggo is spinning its tail so fast that it is about to go into orbit. *Bork!* You have completed the first half of this book. Doggo is incredibly proud of how far you've come. Even if everything still seems shaky, you now have a general idea of the math behind orbital dynamics. This resource isn't going anywhere and you can always return to it if something becomes fuzzy.

Doggo is now flying fast, breaking through the stratosphere. Woof! Before he actually ends up in orbit, Doggo wishes to tell you what is up ahead. We are switching gears from satellite orbits to communications technology. Doggo knows you can do it and has left its favorite bone with you for good luck!

Part II

Satellite Communications

Chapter 6

Link Design & Analysis

6.1   Introduction to Link Design

A communication link is the connection, or path, between a transmitter and a receiver. While transmitters and receivers both use antennas, they have different kinds of amplifiers. Transmitters have high-power amplifiers (HPAs), which ensure the signal's power is sufficiently strong to travel thousands of kilometers to a receiver on the other end of a satellite communications link. If the link is an "uplink", then the transmitter is located on the ground, and the receiver is in space. If the link is a "downlink", then the transmitter is in space, and the receiver is located on the ground! Receivers consist of a low noise amplifier (LNA), which amplifies the faint signal after it has traveled thousands of kilometers from space. A communications system with the uplink, downlink, transmitter, and receiver is shown in Figure 6.1. It is important to note that only the transmit RF chain is depicted at the gateway and only the receive RF chain is depicted at the user terminal.

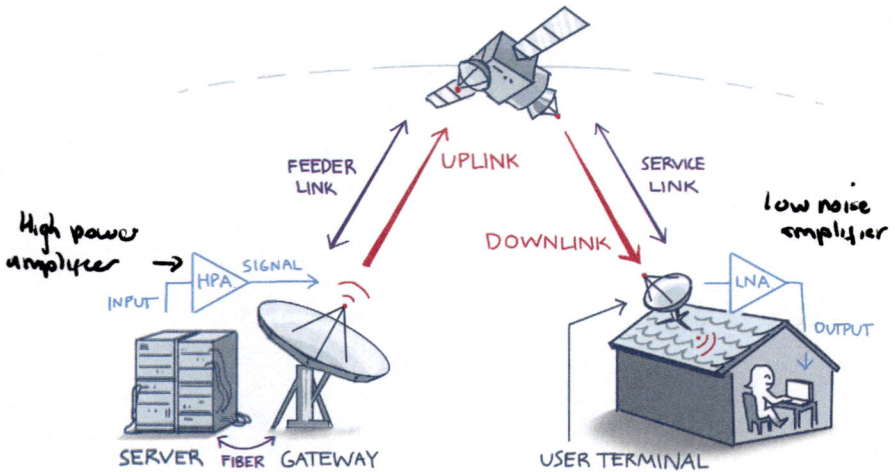

Figure 6.1: Gateway to User Terminal

In the next few chapters, we will build the tools and skills necessary to design a communications link and calculate its performance. We'll approach these topics in the following order:

1. **Transmitters**: How do we transmit, or send, our signal? When discussing transmitters, we will detail the amplification and loss of a signal in the transmission system.

2. **Signal propagation**: How does our signal travel and attenuate through the space between the transmitter and receiver?

3. **Receivers**: How do we receive and process our signal? Like with transmitters, when discussing receivers, we will detail the amplification and loss of a signal in the receiver system.

4. **Overall link performance**: How do we evaluate the performance of our overall link? We will look into several performance metrics, such as total signal strength and data rates.

Before we dive into the technical details, let's start with the bigger picture of frequency allocations and spectrum and take a look at decibels, the unit of signal strength!

6.2 Frequency Allocations and Bandwidth

Table 6.1 provides an overview of the frequency ranges that define several common frequency "bands". You can think of a band as a generally agreed-upon highway or channel on which we can allocate and coordinate traffic. In this case, our traffic is radio waves encoded with information and not cars.

Creating these standards is helpful for many reasons, but let's quickly touch on two. For one, agreed-upon bands allow us to use standardized equipment. Imagine how expensive it would be if you had to custom-design your own communication system from scratch. With standard bands, we can leverage economies of scale for chips and antennas. But you may ask, why draw lines at these specific numbers? If these definitions seem arbitrary, then your intuition is serving you well, because they aren't limited by a mathematical construct, but more which services have been historically operating in the bands!

Our second reason is that bands give names to otherwise abstract concepts, allowing global coordination and easier discussion. While the numbers you see are arbitrary, the allocations are the result of decades of international negotiation and coordination.

Table 6.1: Frequency Allocations

Frequency Range	IEEE Radio Band
30 MHz - 300 MHz	Very High Frequency (VHF)
300 MHz - 3 GHz	Ultra High Frequency (UHF)
1 GHz - 2 GHz	L Band
2 GHz - 4 GHz	S Band
4 GHz - 8 GHz	C Band
8 GHz - 12 GHz	X Band
12 GHz - 18 GHz	Ku Band
18 GHz - 27 GHz	K Band
27 GHz - 40 GHz	Ka Band

The International Telecommunications Union (ITU) is housed under the United Nations (UN) and regulates spectrum globally. Every three to five years, the ITU hosts a World Radio Conference (WRC) [11]. In the periods between the conferences, the ITU organizes working party meetings where nations and corporate entities convene to take on studies for updating the Radio Regulations (RR).

In addition to the ITU, each nation has a governing body responsible for regulating radio communications. In the US, there is the Federal Communications Commission (FCC). In the UK, there is an entity called the Office of Communications (Ofcom). Each nation can define its own unique frequency allocations table, but they often work to align national allocations with the ITU's frequency allocations (codified in Volume 1, Article 5 of the ITU Radio Regulations) [16]. This alignment can enable spectrum allocations to be "globally harmonized", which is helpful for satellite systems that orbit over and operate across many nations. Just imagine if you had to design a global satellite system that had to change frequencies every time you were over a different country! Especially in places like Europe, which has clusters of nations with relatively smaller land mass. Now that would be challenging!

History with Squit!

Changing the frequency over which a satellite operates isn't always as easy as changing the channel of your television. An operational frequency range is chosen far in advance before a satellite is actually built. If satellites didn't use a common set of frequencies then every piece of radio equipment would have to be custom built for each satellite! Squit is not an economist but that would surely cost a lot!

When discussing **bandwidth**, we refer to a chunk of the electromagnetic spectrum, or a "range of frequencies" over which a channel or set of channels operate. One could design a system with five 100 MHz channels to occupy the full 500 MHz bandwidth and describe each channel as having a channel bandwidth of 100 MHz.

Guglielmo Marconi is widely credited as the inventor of the radio [26]. Born in Bologna, Italy, he experimented with wireless communications on his father's estate [26]. After iterating on his design, he was able to transmit a signal across the Atlantic Ocean in 1901, a distance of over 2000 miles [26]. Eight years later, he shared a Nobel Prize for Physics with Karl Ferdinand Braun for his contributions toward the wireless telegraphy [26].

Figure 6.2 displays a schematic of channel bandwidth, B, which is defined between the lower and upper frequency of the channel. The portion between the red curve and the dotted blue "bandwidth" lines is known as the **"roll-off" factor**. Unfortunately, due to imperfections in electronic equipment, it is not possible to transmit at maximum power across the entire channel bandwidth.

To prevent interference into other systems, we often space our channels, like the one shown in Figure 6.2, greater than one bandwidth apart from its neighboring channel to create a guardband.

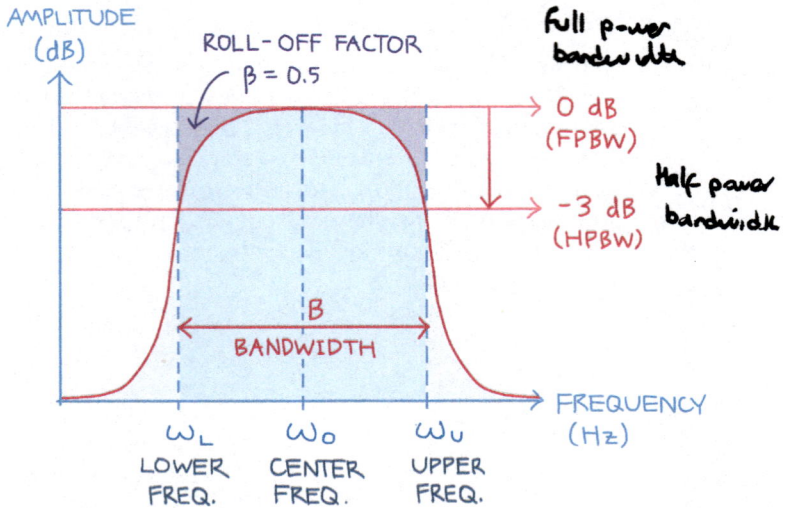

Figure 6.2: Channel Bandwidth and Roll-off

In Figure 6.2, the center frequency of the channel, as well as the full power and half power bandwidth, FPBW and HPBW, respectively, are defined. These terms provide us with insights on the amount of power contained within a channel. In the next section, we will learn more about the decibel (dB) and understand how −3 dB translates to a half.

6.3 Decibels

For a historical timeline of the decibel, which began in the mid-1800s, check out Figure 6.3! In 1858, the first Atlantic telegraph cable was laid, and the original unit of loss, the Miles Standard Cable (MSC), was defined as the amount of loss of a predefined signal over one mile of standard cable [20]. Nearly sixty years later, Bell Telephone Labs replaced the MSC with the Transmission Unit (TU), which would later become the decibel in honor of Alexander Graham Bell [20].

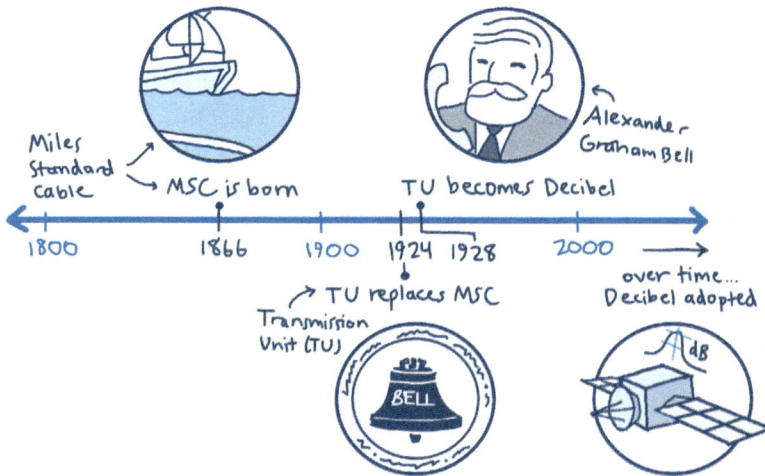

Figure 6.3: Decibel History

You may have heard of decibels in the context of sound and volume, but decibels can measure so much more! Over time, the decibel has become the standard for representing losses and gains throughout a communications system. A way to think of a loss or a gain is essentially a change in the output of a component compared to the input, which we will quantify as a ratio.

Next, we will see how to convert this ratio into decibels using the logarithm $10\log_{10}$ of that ratio. If this is confusing, stay with us for a few more paragraphs! We're going to practice this together too.

To convert a value into decibels, we use a logarithm because it can represent an extensive range of ratios with a simple number (like with scientific notation). For example, a sound wave at 120 dB is a trillion times more intense than 0 dB, which for sound is the typical volume that humans start hearing! Instead of saying a trillion (that's a bunch of zeros!), we can use 120 dB because $10\log_{10}(1E12)$ is 120 dB.

Doggo Help

> Doggo brings you an asterisk! Bork! Bork! Apparently, the last paragraph was not completely truthful. There are decibels that are $20\log_{10}$ of any ratio but these measure magnitude instead of power. You can assume that all of the decibels in this textbook will be $10\log_{10}$.

In short, decibels are used to make working with values spanning orders of magnitude easier. Equation 6.1 shows how to reflect the ratio between a measured power and a reference power in decibels.

$$X = 10 \log_{10} \left(\frac{P_{\text{measured}}}{P_{\text{reference}}} \right) \text{ [dB]} \qquad (6.1)$$

The ratio between a measured and reference power is referred to as a **power ratio** and is unitless because the numerator and denominator have the same units. Because decibels are technically unitless (similarly to radians), they are often measured with respect to a reference unit such as W for Watts or m for milliwatts. This letter tells you that the measured signal is some factor larger than *one* of those units. For instance, a value of 10 dBW denotes a measured power of ten times larger than 1 Watt. A value of 3 dBm denotes a measured power that is two times larger than one milliwatt.

These reference letters are important. Take a moment to review the bulleted list of common reference units:

- **dBW:** ratio for which the reference is 1 W

- **dBm:** ratio for which the reference is 1 mW

- **dBc:** ratio for which the reference is the carrier

- **dBi:** ratio for which the reference is an isotropic antenna

- **dBd:** ratio for which the reference is a dipole antenna

- **dB-Hz:** ratio for which the reference bandwidth is 1 Hz

- **dB/K:** ratio which relates gain to system noise temperature

Let's take a moment to practice this together.

Example with Sat!

Convert 100 W into decibels.

Solution:
We can use Equation 6.1 to determine what 100 W, a non-logarithmic power ratio, is in decibel space, assuming a reference power of 1 W.

$$X \text{ dB} = 10 \log_{10} \left(\frac{P_{\text{measured}}}{P_{\text{reference}}} \right)$$

$$= 10 \log_{10} \left(\frac{100 \text{ W}}{1 \text{ W}} \right)$$

$$= 20 \text{ dBW}$$

Table 6.2 can serve as a guide for linear power values that will be used often and that are worth memorizing.

Table 6.2: Key Decibel Values

Linear (X)	Base 10 Logarithm	dB = $10\log_{10}(X)$
$\frac{1}{1000}$	10^{-3}	-30
$\frac{1}{100}$	10^{-2}	-20
$\frac{1}{10}$	10^{-1}	-10
$\frac{1}{2}$	$10^{-0.3}$	$\sim$-3
1	10^0	0
2	$10^{0.3}$	~ 3
3	$10^{0.5}$	~ 5
5	$10^{-0.7}$	~ 7
10	10^1	10
100	10^2	20
1000	10^3	30

To convert a value from decibel space (which is logarithmic) back into linear (power ratio) space, we can use Equation 6.2:

$$\text{Power Ratio } = 10^{\frac{X \text{ [dB]}}{10}} \text{ [unitless]} \tag{6.2}$$

Now, let us look at an example of converting from dB to power ratio.

Example with Sat!

Consider an amplifier with a gain of 30 dB. Calculate the power ratio of the amplifier.

Solution:
We can use Equation 6.2 to determine the power ratio:

$$\text{Power Ratio } = 10^{\left(\frac{X \text{ dB}}{10}\right)}$$
$$= 10^{\left(\frac{30 \text{ dB}}{10}\right)}$$
$$= 10^3 = 1000$$

While it's important to know how to compute this manually, you can also refer back to the Key Decibel Values table as a reference and move from right to left.

Note that because decibels utilize logarithms, arithmetic is slightly different than when we work in linear space. To summarize and explain further:

- Addition in log space (dB) becomes multiplication in linear space (power ratios).

$$\log(A) + \log(B) + \log(C) = \log(A \times B \times C) \tag{6.3}$$

- Subtraction in log space (dB) becomes division in linear space (power ratios)

$$\log(A) - \log(B) = \log\left(\frac{A}{B}\right) \tag{6.4}$$

In a previous example with Sat, we saw that 100 W is equivalent to 20 dBW. To solve the equation $10 \log_{10}(100 \text{ W})$, you may have used a program like Excel. Another approach is to factor 100 W into key decibel values provided in Table 6.2. For example, 100 can be factored into 10 * 10, or 25 * 4, which can be further factored into 5 * 5 * 2 * 2. Using the properties of logarithms, multiplication becomes addition, but the key is to convert each of our factors into decibels before we add. Therefore, 10 * 10 would become 10 + 10, or if we factored using the second approach, 5 * 5 * 2 * 2 would become 7 + 7 + 3 + 3 = 20 dB.

6.4   Problem Set 4: Link Design Intro

Problem	Topic	Points
1	Decibel Worksheet	10
2	Decibel Reflection	3
3	Bandwidth Review	3
4	Gain of Two Amplifiers in Series	6
5	Transmitter Performance	6
Total:		28

Exercise 6.1

(10 points) Complete the following decibel conversion table.

Ratio	Ratio factored in linear space	Ratio in logarithmic space	dB
20	2×10	$3 + 10$	13
25	$100/(2 \times 2)$	$20 - (3 + 3)$	14
500			
0.4			
			-3
			-13
			-30
20 W			(in dBm)
50 mW			(in dBm)
50 mW			(in dBW)

Exercise 6.2

(**3 points**) Compare your table and approach with a partner. If you're learning this on your own, see if you can teach a friend how to convert in and out of decibels and have them complete this exercise.

1. What do you observe when you compare your solutions?

2. Did you approach computations differently?

3. Which conversions were most challenging and most intuitive?

Exercise 6.3

(**3 points**) Let's take a look at Figure 6.2 again to better understand bandwidth. For this question, you might have to do a little bit of external research to find the correct answer.

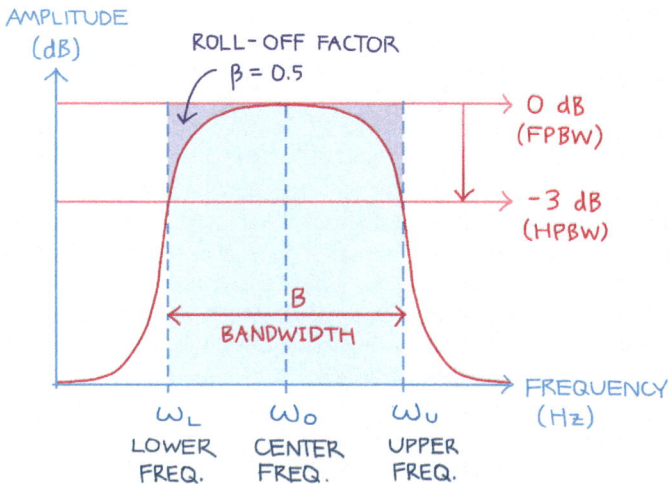

1. The channel is shaped like a "raised-cosine filter." What is the raised-cosine filter and why is it called that?

2. What does the signal look like when our roll-off factor, β, is zero? Is there a name for this type of signal?

3. Looking at the figure, in which $\beta = 0.5$, over what frequency is the full power bandwidth (FPBW) defined? How do you know? Over what frequencies is the HPBW defined?

4. Assume that, for the above channel, $\beta = 0$. What is the frequency range of FPBW? What about HPBW?

Exercise 6.4

(**6 points**) Suppose we have two amplifiers in series; one has a gain of 100, and the other has a gain of 2. Since the amplifiers are in series, we can compute the total gain of the system by multiplying the gains together.

1. What is the total gain?

2. What is the total gain in decibels? First, convert the two individual gains into decibels.

Exercise 6.5

(**6 points**) Skipping ahead slightly, let us consider a transmitter that propagates signals at a power, P, of 100 W. The gain, G, of the transmitter's antenna, which essentially tells us how much the signal is amplified at transmission, is 1.2. Finally, parts of the signal can get lost during transmission. The loss, L, of this transmitter is 25 in linear space.

1. To compute the transmitter's performance, multiply our power level by our gain and divide it by our loss, i.e., $\frac{P*G}{L}$. What is value of this metric in linear space?

2. Now perform the calculation with decibels. Convert P, G, and L into decibels and use decibel arithmetic to calculate the performance. Note that for measurements in dBW and dB, X dBW $+ Y$ dB $= (X + Y)$ dBW. It's often wise to confirm whether you compute the same answer when converting it back to linear space.

Chapter 7

Transmitters (Tx)

In this chapter, we will cover the transmission system, which consists of a power amplifier and an antenna connected by a transmission line, which feeds the signal into the antenna for transmission. Figure 7.1 provides a schematic of each of these components. The transmission line is also referred to as the feeder.

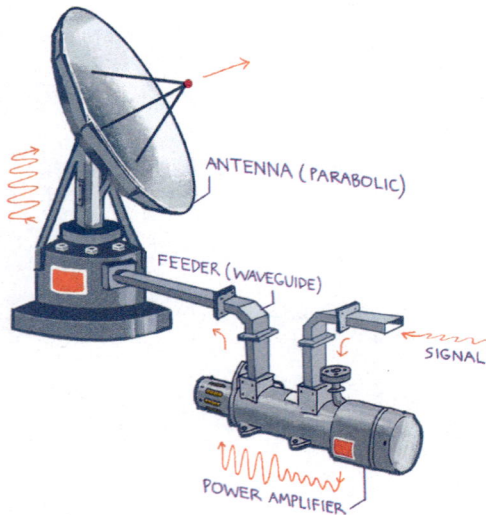

Figure 7.1: Transmitter

For transmission, data is first encoded in an electromagnetic wave, called a *signal*. This data signal is then fed into a power amplifier to boost its strength. Next, the high-power signal is provided to the antenna, which radiates the signal out into space. We care about the strength of the signal throughout this entire process, so in this chapter, we'll learn about the various parts (e.g., power amplifiers, feeders, etc.) of the transmitter system and then shift into antenna theory covering antenna gain, beamwidth, pointing, and polarization.

> **History with Squit!**
>
> Arghhhh! If we only care about the strength of the signal, why don't they increase the power to kingdom come?! Well ye mates, what is a signal for you is noise for others, so the louder (stronger) your signal, the more difficult it will be for everyone around you to receive their own signal. This phenomenon is called interference. Regulatory bodies, like the FCC and ITU, work closely with their license applicants to ensure that harmful interference is mitigated amongst systems.

7.1 Power Amplifiers

Both transmitters and receivers consist of at least one power amplifier (PA), which amplifies the power of an input signal. In a transmitter, PAs take in a low power RF signal and output a higher power version, which is then fed into an antenna. The output power is a function of the efficiency of the PA, which is frequency dependent. Increasing PA efficiency is one of the fundamental challenges in satellite communications that engineers are actively working to address; if we can increase the efficiency of our power amplifiers, some of which have an efficiency of approximately 0.3 - 0.45, then the overall cost of providing service via satellite networks will decrease!

Example with Sat!

If the transceiver (a radio that can both transmit and receive) has an input signal to the PA with a power of 20 W with a PA efficiency of 0.3, what is the RF output power of the PA in W, dBW, and dBm?

Solution:

To compute the power in Watts, we multiply 20 by 0.3 and find a result of 6 W. If we want to convert this value into decibels, we compute $10 \log_{10}(6 \text{ W}) = 7.78$ dBW. Lastly, since there are 1,000 milliwatts (mW) in a Watt (W), we add 30 dB to 7.78 dBW, which comes to 37.78 dBm.

The two primary types of amplifiers, illustrated in Figure 7.2, are Solid State Power Amplifiers (SSPAs) and Travelling Wave Tube Amplifiers (TWTAs) [22]. SSPAs are integrated circuits that use networks of transistors to amplify low power signals, whereas TWTAs are specialized vacuum tubes that use electron beams for amplification [22].

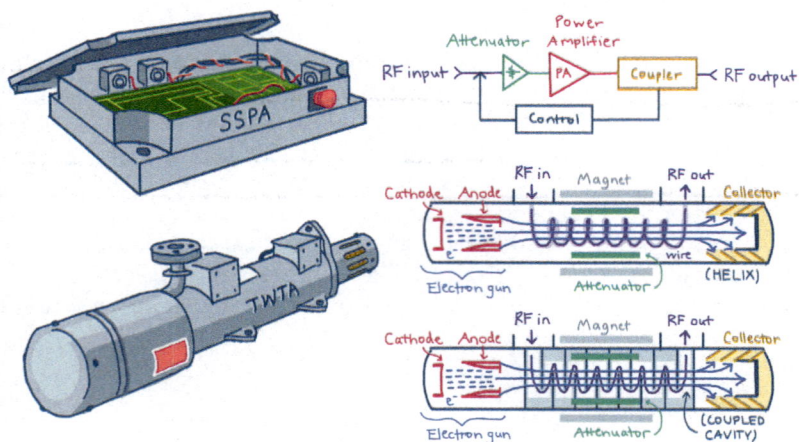

Figure 7.2: High Power Amplifiers: SSPAs and TWTAs

TWTAs are further classified as either *helix* or *coupled cavity*. In helix TWTAs, a radio wave is sent through a helix-shaped wire, and a beam of electrons is shot through the center of the helix. The beam-radio interaction amplifies the signal [22]. Coupled cavity TWTAs overcome this problem by replacing the wire with a series of resonant cavities that guide the radio waves helically. Again, the beam-radio interaction amplifies the signal. These TWTAs are only narrowband but can amplify to extremely high powers [22]. Because of their physical design, TWTAs tend to be relatively large and fragile. They also tend to consume vast amounts of power.

SSPAs, by contrast, rely on modern integrated circuits. Through networks of resistors, capacitors, inductors, and power transistors, SS-PAs amplify signals on circuit boards that are compact and mechanically sturdy. Due to their compact size and lower power consumption, SS-PAs are the primary type of power amplifiers used particularly on small satellites today [22].

7.2 Feeders

In between the amplifier and the antenna is a transmission line, or feeder, which carries signals from the amplifier to the antenna. Transmission lines often take the form of waveguides or coaxial cables in radio frequency (RF) systems. Ideal feeders transfer the power between the PA and the antenna without decreasing signal strength; however, in the real world, power degradation, or loss, occurs as the signal moves through the feeder. In most link assessments, we assume a feeder loss between 0.5 dB and 1.5 dB. In reality, feeder loss must be measured with a power meter that inputs a signal of known strength through a feed line and determines the difference at the output of the feed.

7.3 Antennas

Antennas are electrical conductors, which are usually pieces of metal that are tuned to resonate at specific electromagnetic frequencies, enabling them to convert electrical power into electromagnetic waves that can travel through free space and vice versa. Because antennas convert power between electromagnetic waves and electrical current, they are considered *transducers*. It is common to talk about transmit, T_x or Tx, and receive, R_x or Rx, antennas as though they are different, but in practice, a single antenna is a "reciprocal device" and can perform both functions. The labeling is useful to define in which direction an antenna is converting electrical power:

- Transmit (T_x) antennas accept electrical current and radiate, or couple, that power into free space. During transmission, a guided wave of electrical power is applied to the conductor. As this guided wave resonates inside the conductor, electromagnetic waves are released into free space in a process called *propagation*.

- Receive (R_x) antennas do the opposite. An electromagnetic wave traveling through space is absorbed by the antenna, inducing an alternating current in the conductor. The antenna efficiently concentrates this current into a guided wave, and transmission lines carry the wave from the conductor to a receiver.

Antenna theory describes what happens during this process, both inside the antenna and in the surrounding free space.

Doggo Help

Antennas may seem magical, like a talking dog, but you can think of them as tuning forks that transmit electromagnetic waves instead of sound! Remember how larger tuning forks produce lower notes (low frequencies) and smaller tuning forks have a higher pitch (high frequencies)? Well, the same basic sizing is true for antennas.

Converting electrical current into invisible electromagnetic waves is, admittedly, an abstract concept. To better conceptualize what is happening here, imagine a transmitting antenna and draw a sphere around it. Consider the antenna at the center of the sphere. The power emitted by the antenna must pass through the surface of this sphere as it radiates out into space. By measuring the amount of power passing through portions of the sphere, we can define power densities for various parts of the antenna pattern. Figure 7.3 illustrates spherical coordinates with elevation and azimuth angles corresponding to longitude or azimuth, θ, and latitude or elevation, ϕ, respectively.

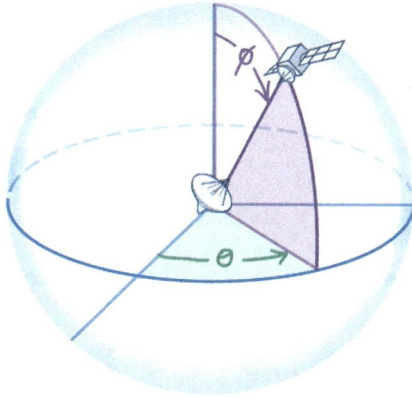

Figure 7.3: Standard Spherical Coordinate System

Two particularly important concepts in antenna theory are the total and partial surface areas of a sphere. The total surface area of a sphere is given by $4\pi r^2$. Portions of spherical surface area are often measured with the SI unit steradians. For a circle of radius, r, a **radian** is the angle swept out with an arc of length, r, and is a unitless quantity. For a sphere of radius, r, a **steradian** is the angle of a cone that "slices" out an area of r^2 on the sphere's surface. This unitless angle is called a *solid angle* and is often denoted by capital omega, Ω. Similar to how a full circle contains 2π radians, and has a circumference of $2\pi r$, a full sphere contains 4π steradians and has a surface area of $4\pi r^2$.

An **isotropic** antenna is an ideal antenna that radiates energy equally in all directions, creating a spherical antenna pattern. This type of antenna also converts all of its power into radio waves with 100% efficiency. Although isotropic antennas do not exist, they are a valuable, theoretical concept for defining gain.

7.4   Gain

Unlike isotropic antennas, realizable antennas are designed to transmit or receive power in a specific direction. Gain is the ratio between the power radiated per unit solid angle by an antenna and the power radiated by an isotropic antenna fed with the same power. Simply put, gain compares the actual amount of power transmitted by your antenna to the power that would be transmitted by an isotropic antenna fed with the same power. If this explanation isn't clear, try to stick with us, as we will think about this concept differently right after a little bit of math.

Gain can be expressed mathematically with the following equation:

$$\text{Gain} = \frac{\text{Actual power emitted in a given direction}}{\text{Theoretical power emitted by an isotropic antenna}} \qquad (7.1)$$

Because gain is a unitless ratio, it is often expressed in dBi; the lowercase 'i' denotes that the power ratio is defined with respect to an isotropic antenna. Isotropic antennas have a gain of 0 dBi for the following reason:

$$\begin{aligned}
\text{Gain}_{\text{iso}} &= 10 \log_{10} \left(\frac{\text{Actual power}}{\text{Theoretical power}} \right) \\
&= 10 \log_{10} \left(\frac{\text{Isotropic power [W]}}{\text{Isotropic power [W]}} \right) \\
&= 10 \log_{10}(1) \\
&= 0 \text{ [dBi]}
\end{aligned}$$

For a different way to consider gain, imagine a light bulb in the middle of an empty room. Any location in the room sees the same amount of brightness from the bulb, making this an *isotropic* light bulb with uniform brightness, B.

Now, imagine painting half of the light bulb in silver paint so that all of the light from that half of the bulb is reflected back out the other side. One half of the room is receiving no light, while the other half is receiving twice as much! The light bulb is still emitting the same total amount of light, but the brightness on the lit half of the room is $2B$. Since gain is the ratio between the two brightnesses, the gain of the half-painted light bulb is $2B/B = 2$, or by convention, $10 \log_{10}(2) = 3$ dBi. Recall that a factor of two in linear space is an addition of $+3$ dB in logarithmic space.

Now, what if we paint $3/4$ of the light bulb with silver paint so that all of the light is reflected out into only $1/4$ of the room? Although $3/4$ of the room will be dark, the bright quarter of the room will see $4B$ of brightness! The gain here is $10 \log_{10}(4B/B) = 6$ dBi. To sanity check that we did our math correctly, we can notice that the $3/4$ light bulb is twice as bright as the half-painted light bulb, a factor of two, and the decibel value increased by $+3$ dB. We know that 2x in linear space is $+3$ dB in logarithmic space, so we did our math correctly!

Although antennas aren't light bulbs, it's a decent approximation to think of an antenna as a sophisticated light bulb engineered to transmit concentrated amounts of light in particular directions.

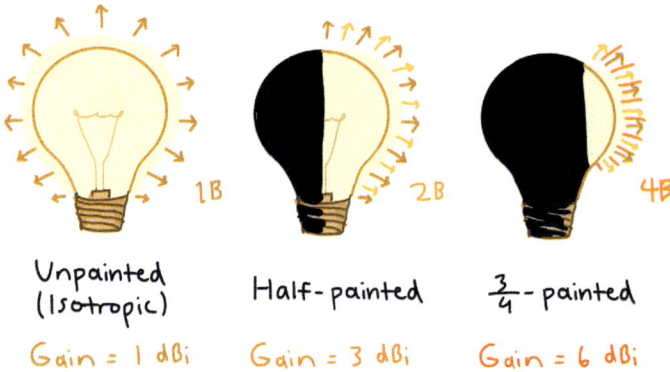

Figure 7.4: Light Bulb Representation of Gain

The maximum gain of an antenna is related to its physical geometry and the signal frequency. An expression for maximum gain in linear space is given by:

$$G_{\text{max}} = \left(\frac{4\pi}{\lambda^2}\right) A_{\text{eff}} \ [\text{unitless}] \tag{7.2}$$

where A_{eff} is the effective aperture area of the antenna in squared meters, and λ is the wavelength in meters. It may be useful to note that the speed of light, c, the wavelength, λ, and the frequency, f, of a signal are related by $c = \lambda f$. Because antenna gain depends on wavelength, and wavelength and frequency are related by the speed of light, antenna gain also varies with frequency. For an antenna with a circular aperture or reflector of diameter, d, the effective aperture is calculated with Equation 7.3.

$$A_{\text{eff}} = \eta A_{\text{circle}} = \frac{\eta \pi d^2}{4} \ [\text{m}^2] \tag{7.3}$$

where η is the efficiency of the antenna and $A_{\text{circle}} = \frac{\pi d^2}{4}$. Substituting Equation 7.3 into Equation 7.2 yields the maximum gain equation:

$$G_{\text{max}} = \eta \left(\frac{\pi d}{\lambda}\right)^2 \ [\text{unitless}] \tag{7.4}$$

To convert Equation 7.4 to decibels, either of the following equations may be used:

$$G_{\text{max,dB}} = 10 \log_{10} \left[\eta \left(\frac{\pi d}{\lambda} \right)^2 \right] \quad [\text{dBi}] \qquad (7.5)$$

which multiplies the linear terms prior to converting to the logarithmic space, or the second approach, which breaks the product of the efficiency and aperture area into a summation of two logarithmic quantities.

$$G_{\text{max,dB}} = 10 \log_{10}(\eta) + 20 \log_{10} \left(\frac{\pi d}{\lambda} \right) \quad [\text{dBi}] \qquad (7.6)$$

As always, let's take a moment to practice what we have just learned about gain.

Example with Sat!

Solve for the gain of an antenna with 65% efficiency ($\eta = 0.65$) and a diameter of 4 meters. Compare the gains at frequencies of 1 GHz and 12 GHz.

Solution:

To compute the maximum gain, we plug in the given values into the gain equation and convert them to decibels:

$$G_{\max} = \eta \left(\frac{\pi d}{\lambda} \right)^2$$

$$G_{\max} = 10 \log_{10}(\eta) + 20 \log_{10} \left(\frac{\pi d}{\lambda} \right)$$

$$G_{\max} = 10 \log_{10}(\eta) + 20 \log_{10} \left(\frac{\pi d f}{c} \right)$$

Next, we compute the gain at 12 GHz:

$$G_{12\text{GHz}} = 10 \log_{10}(0.65) + 20 \log_{10} \left(\frac{4\pi(12 E9 \text{ Hz})}{3 E8 \text{ m/s}} \right)$$

$$= 52.2 \text{ dBi}$$

Then, we compare the gain of the system operating at 12 GHz to the gain at 1 GHz:

$$G_{1\text{GHz}} = 10 \log_{10}(0.65) + 20 \log_{10} \left(\frac{4\pi(1 E9 \text{ Hz})}{3 E8 \text{ m/s}} \right)$$

$$= 30.6 \text{ dBi}$$

Ultimately, when we compare these two gains, we see that the gain at 12 GHz is greater than the gain at 1 GHz. An antenna at a lower frequency, and therefore longer wavelength, will have less gain than an antenna of the same efficiency and aperture area at a higher frequency.

7.5   Radiation Patterns

Radiation patterns are representations of the variation of antenna gain with direction. An omnidirectional antenna is often used for telemetry and telecommand (TT&C), as it radiates power uniformly in all directions in one plane. Figure 7.5 shows radiation patterns for an isotropic, omnidirectional, and directional antenna.

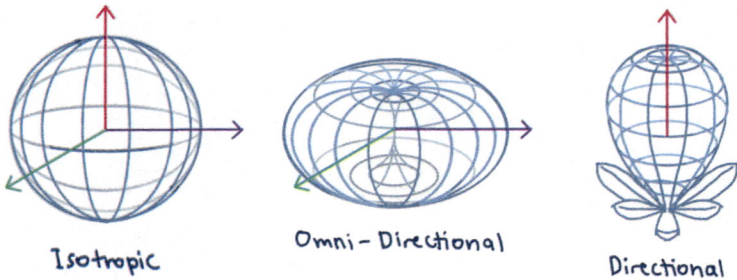

Figure 7.5: Different Types of Radiation Patterns

Figure 7.6 shows common types of antennas used in satellite communications.

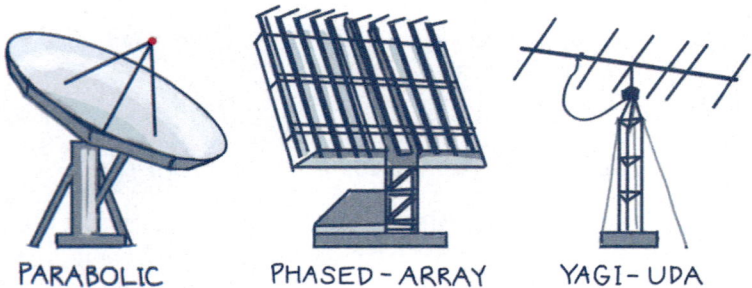

Figure 7.6: Different Antenna Types

Broadband constellations primarily utilize phased array and parabolic antennas, whereas lower frequency amateur radio systems are more likely to use Yagi Udas.

7.6 Beamwidth

As we have seen, antennas have a maximum, or peak, radiation direction generally along the axis of symmetry of the radiation pattern, often called *boresight*. To quantify the width of the high-gain region, we define a quantity called beamwidth.

The **beamwidth** is the angle between the peak gain and a given decrease in gain. This "given decrease" may be any value, but we often use -3 dB to denote half the maximum beam strength. Industry refers to this as "the 3 dB beamwidth", or the half power beamwidth (HPBW). Figure 7.7 shows a polar representation of a gain pattern that has a "main lobe" of high gain and several off-axis "sidelobes". The boresight of this antenna is down the middle of the main lobe, and the 3 dB beamwidth, $\theta_{3\text{dB}}$, is the angle between the boresight and the location on the main lobe where the signal strength is half the peak level.

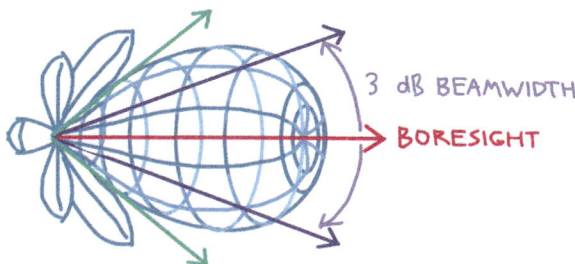

Figure 7.7: Antenna Radiation Patterns

The 3 dB beamwidth, measured in degrees, is related to the ratio of wavelength to antenna diameter by a scaling coefficient, k:

$$\theta_{3\text{dB}} = k\left(\frac{\lambda}{d}\right) = k\left(\frac{c}{fd}\right) \quad [\text{unit of } k] \tag{7.7}$$

Because the wavelength and diameter both have units of meters, the ratio between them is unitless, and k drives the units of $\theta_{3\text{dB}}$. A value of $70°$ is commonly used for parabolic antennas, but you may also see a value of $51°$. We'll stick to a coefficient of $70°$ in this text:

$$\theta_{3\text{dB}} = 70\left(\frac{\lambda}{d}\right) = 70\left(\frac{c}{fd}\right) \quad [\text{degrees}] \tag{7.8}$$

The primary takeaway is that beamwidth provides a metric for how a signal's power level changes over various pointing angles. To determine how this affects the strength of the overall link, we must first figure out how accurately our system can point and steer the antenna.

Example with Sat!

Compute the 3 dB beamwidth of a three-meter antenna that transmits at Ku-band (14.0 GHz).

Solution:
To compute the 3 dB beamwidth, we utilize Equation 7.8.

$$\theta_{3\text{dB}} = 70 \left(\frac{c}{fd} \right)$$

$$= 70 \left(\frac{3E8 \text{ m/s}}{(14E9 \text{ Hz})(3 \text{ m})} \right)$$

$$= 0.5°$$

The solution has units of degrees because of the leading co-efficient.

7.7   Pointing Loss

Pointing loss is the decrease in signal power due to misalignment between the main beam and the intended boresight. In other words, pointing loss is the difference between the maximum gain of the antenna and the actual gain in the real-world system.

A satellite's orientation, also called its attitude, is controlled by an "attitude control system", or ACS, made up of small jets, spinning disks, and electromagnets that keep the satellite oriented within a certain error. Pointing accuracy, measured in degrees, is the maximum error expected between a satellite's actual and expected orientation. The pointing loss, L_{Pointing}, can be calculated in decibels from pointing accuracy, θ_{Pointing}, and the half power beamwidth, $\theta_{3\text{dB}}$, as shown in Equation 7.9:

$$L_{\text{Pointing}} = 12 \left(\frac{\theta_{\text{Pointing}}}{\theta_{3\text{dB}}} \right)^2 \text{ [dB]} \qquad (7.9)$$

While the ratio between the pointing accuracy and beamwidth is calculated in linear space, the output of this equation is in decibels. In this next example, we will see the impact of frequency on pointing error.

Example with Sat!

Compute the pointing loss for a 3-meter antenna that transmits at the Ku-band (14.0 GHz) if it has a 0.5° pointing accuracy. How is the pointing loss impacted if we transmit from the same system at V-band (50.4 GHz)?

Solution:
In the previous example, we calculated the 3 dB beamwidth of a 3-meter antenna to be 0.5°. Using this, along with the given pointing accuracy of 0.5°, we can compute the pointing loss:

$$L_{\text{Pointing}} = 12 \left(\frac{\theta_{\text{Pointing}}}{\theta_{\text{3dB}}} \right)^2$$

$$= 12 \left(\frac{0.5°}{0.5°} \right)^2$$

$$= 12.0 \text{ dB}$$

If we transmit from the same system at V-band (50.4 GHz) and recompute the 3 dB beamwidth, we find that the beamwidth shrinks from 0.5° to 0.14°:

$$\theta_{\text{3dB}} = 70 \left(\frac{c}{fd} \right)$$

$$= 70 \left(\frac{3E8 \text{ m/s}}{(50.4E9 \text{ Hz})(3 \text{ m})} \right)$$

$$= 0.14°$$

Next, we can use this beamwidth to calculate L_{Pointing}:

$$L_{\text{Pointing}} = 12 \left(\frac{\theta_{\text{Pointing}}}{\theta_{\text{3dB}}} \right)^2$$

$$= 12 \left(\frac{0.5°}{0.14°} \right)^2$$

$$= 153.06 \text{ dB}$$

7.8   Polarization

The radiated electromagnetic wave contains an electric, $\vec{E}$, and a magnetic field, $\vec{B}$, component. These components are orthogonal to each other and perpendicular to the direction of propagation of the wave. For example, if a wave's $\vec{E}$ field component is in a vertical plane, then the $\vec{B}$ field component is in the horizontal plane, and the line formed where these two planes intersect is the direction of propagation. Polarization describes the orientation of the electric field component. An $\vec{E}$ field component that oscillates in a fixed vertical plane is **vertically polarized**, while an $\vec{E}$ field component that oscillates in a fixed horizontal plane is **horizontally polarized**. Examples of both forms of linear polarization are depicted in Figure 7.8.

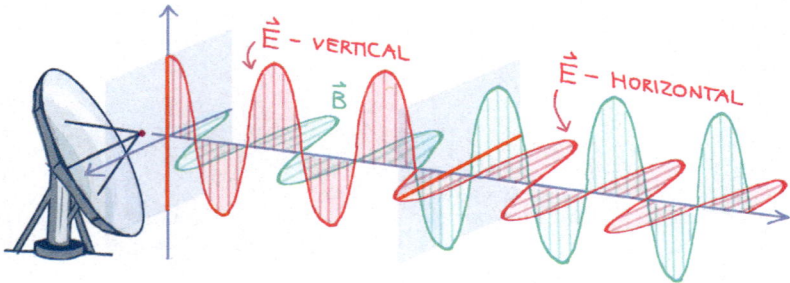

Figure 7.8: Linear Polarization of the Electromagnetic Wave

Doggo Help

You have probably heard of polarization when it comes to sunglasses, but what does it mean when something is horizontally versus vertically polarized? Well, for sunglasses, a horizontally polarized lens will only allow horizontal light to pass through, blocking vertically polarized light and vice versa! If you ever shift around a polarized pair of sunglasses, you probably have noticed that shifting it in one direction makes the lens darken instead of clear. That's the polarization at work!

In some cases, the $\vec{E}$ field's plane of oscillation does not remain fixed vertically or horizontally but instead rotates around some axis. If this is the case, then the tip of the $\vec{E}$ field vector will trace out an ellipse when projected in time onto a plane perpendicular to the direction of propagation. Such waves are said to have **elliptical polarization**. Sometimes, the $\vec{E}$ field plane may rotate around the axis of propagation, represented by our perpendicular $\vec{E}$ and $\vec{B}$ planes rotating like a pinwheel around the line of intersection.

When this occurs, the tip of the $\vec{E}$ field vector will trace out a circle on the plane perpendicular to the direction of propagation. These waves are said to have **circular polarization**, and are a special case of elliptically polarized waves. In a circularly-polarized antenna, the plane of polarization rotates making one full revolution during each wavelength. Waves with $\vec{E}$ fields that rotate clockwise with respect to the direction of propagation are **right-handed** waves, whereas counterclockwise rotations are **left-handed**. Point your thumb along the direction of propagation and let your fingers curl to see why this naming convention works. Figure 7.9 illustrates right-handed and left-handed circular polarization.

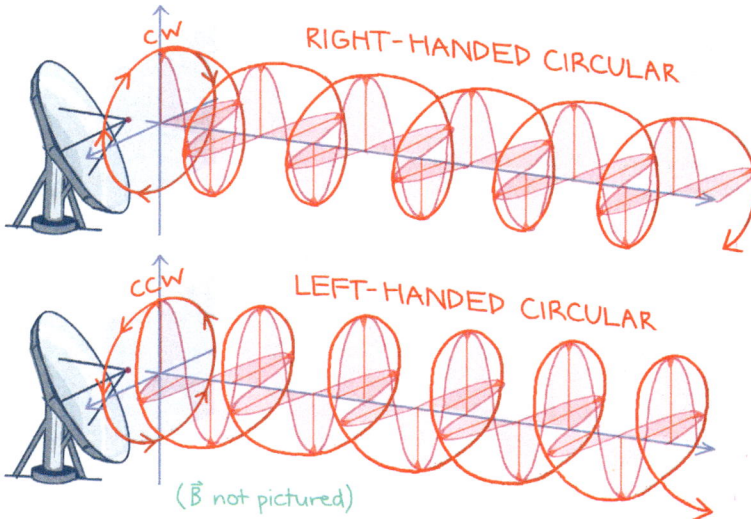

Figure 7.9: Circular Polarization of the Electromagnetic Wave

The polarization of a wave is fully defined by:

- Direction of rotation: Vertical, Horizontal, Right-hand (clockwise) or Left-hand (counter-clockwise) with respect to the direction of propagation

- Axial Ratio (AR): Ratio of E_{max} to E_{min}, or the major / minor axes of the ellipse. When the ellipse is a circle, the axial ratio is 1, or 0 dB, and the polarization is said to be circular. When the ellipse reduces to one axis (infinite AR), the $\vec{E}$ field maintains a fixed direction, and polarization is linear.

In radio systems, when the electric field is perpendicular to the Earth's surface, it is vertically polarized. Automobile whip antennas configured for amplitude-modulated (AM) radio often use vertical polarization [2]. When the electric field is parallel (or tangent) to the Earth's surface, the wave is horizontally polarized. Television transmissions in the United States tend to use horizontal polarization [5]. Geostationary satellites use both horizontal and vertical polarizations. Most non-geostationary (NGSO) satellites deploy circularly polarized systems, and operate using both right-hand circular polarization (RHCP) and left-hand circular polarization (LHCP) simultaneously!

Two waves are in **orthogonal polarizations** if their polarization describes identical ellipses in opposite directions. LHCP is orthogonal to RHCP, and vertical is orthogonal to horizontal. Systems can utilize orthogonal polarizations to transmit or receive at the same frequency at the same time between two locations. This property is known as frequency reuse via orthogonal polarizations, and is a method for efficiently utilizing spectrum.

Polarization mismatch occurs when the receiving antenna is not oriented with the same polarization of the transmitted wave. Antenna polarization alignment is one cause of mismatch, but natural effects can also cause mismatch. Propagation through the atmosphere, for instance, can change circular into elliptical polarization. One can calculate polarization mismatch with the following equation:

$$L_{\text{Polarization}} = -20 \log_{10} \cos \psi \ [\text{dB}] \tag{7.10}$$

where ψ is the angle between antenna polarization planes. Generally, we assume a small value for polarization mismatch, like 0.2 dB. That being said, there are a few special cases when this doesn't hold [2]:

- Linearly (V or H) polarized antenna transmits to a receive antenna that is linearly polarized in the **opposite** direction, the polarization mismatch will result in a $-\infty$ dB, or complete, loss.

- Linearly (V or H) polarized antenna transmits to a receive antenna that is circularly polarized (RHCP or LHCP), the polarization will result in a 3 dB loss, or a factor of 1/2.

- Circularly (RHCP or LHCP) polarized antenna transmits to a receive antenna that is circularly polarized in the **opposite** direction, the polarization will result in a $-\infty$ dB, or complete, loss.

A more advanced link budget may also account for losses due to the utilization of both polarization, which can lead to cross-polarization isolation (XPI), cross-polarization discrimination (XPD). However, we will not address XPI or XPD in this text.

7.9   Effective Isotropic Radiated Power (EIRP)

A transmitter's performance is evaluated with a metric called **effective isotropic radiated power (EIRP)**, measured in dBW. Conceptually, EIRP measures how much power *could effectively* make it to the receiver. To calculate EIRP, we'll start with the amount of raw power input to the receiver, multiply by the antenna gain, and then divide out losses due to pointing error, polarization mismatch, and feeder lines. Sounds like a lot? Don't worry, decibels make this easier! Multiplication and division in linear space are addition and subtraction in logarithmic or decibel space, so calculating EIRP becomes as easy as adding each of the decibel terms together:

$$EIRP = P_{\text{Tx}} + G_{\text{Tx}} - L_{\text{Pointing}} - L_{\text{Polarization}} - L_{\text{Feeder}} \text{ [dBW]} \quad (7.11)$$

In this equation, P_{Tx} is the amplifier power, G_{Tx} is the gain of the transmitter antenna, and the three L terms represent each corresponding loss. Remember, each term must first be converted to decibels before you can add and subtract them!

It is also important to remember that you can add any dB term to another dB term as long as they share the same scale, or fundamental order of magnitude. However, you cannot add two dB terms that were converted into dB from different scalings of the same underlying unit. For instance, you cannot add together a power measured in watts (dBW) and a power measured in milliwatts (dBm). Adding decibel terms with these units would correspond to trying to multiply a power in watts by a power in milliwatts, which can't be done. In contrast, it is acceptable to add gain (dBi) to standard power (dBW).

Example with Sat!

What is the EIRP (measured in dBW and W) of a 2.4 m gateway antenna with 57 dBi of gain, 1.5 dB of feeder loss, and 4 W of RF Output Power?

Solution:

To compute EIRP, we will need to convert all values to decibels, so that we can add or subtract them.

$$
\begin{aligned}
\text{EIRP} &= P_{\text{Tx}} + G_{\text{Tx}} - L_{\text{Feeder}} \\
&= 10 \log_{10}(4 \text{ W}) + 57 \text{ dBi} - 1.5 \text{ dB} \\
&= 6 \text{ dBW} + 57 \text{ dBi} - 1.5 \text{ dB} \\
&= 61.5 \text{ dBW}
\end{aligned}
$$

Converting 61.5 dBW into Watts produces 1419254 W.

Example with Sat!

What is the EIRP of a 0.5 m User Terminal (UT) with 36 dBi of gain, 0.6 W of RF Output Power, and 1 dB of feeder loss?

Solution:

Similarly, we will need to convert non-logarithmic values into decibels prior to computing EIRP.

$$
\begin{aligned}
EIRP &= P_{\text{Tx}} + G_{\text{Tx}} - L_{\text{Pointing}} - L_{\text{Polarization}} - L_{\text{Feeder}} \\
&= 10 \log_{10}(0.6 \text{ W}) + 36 \text{ dBi} - 0 \text{ dB} - 0 \text{ dB} \\
&\quad - 1 \text{ dB} \\
&= -2.2 \text{ dBW} + 36 \text{ dBi} - 0 \text{ dB} - 0 \text{ dB} - 1 \text{ dB} \\
&= 32.8 \text{ dBW}
\end{aligned}
$$

7.10 Problem Set 5: Transmitters

Problem	Topic	Points
1	Conceptual Transmitter Questions	4
2	RF Output Power Given EIRP and Gain	3
3	Parabolic Dish Parameters given Gain	3
4	Transmitter Performance and Characterization	3
5	EIRP and Polarization Mismatch	3
Total:		16

Exercise 7.1

(**4 points**) When answering the following questions conceptually, target an audience of someone with a limited technical background.

1. What is an isotropic antenna? How is it different from an omnidirectional antenna?

2. What is antenna gain, and what formula do we use to calculate gain?

3. What is polarization?

4. **Bonus:** What are Maxwell's equations, and why are they important in the context of antenna theory? P.S. This may require independent research.

Exercise 7.2

(**3 points**) Given a Satellite Transmitter with 40 dBW EIRP, a gain of 30 dBi, and T_x feeder loss of 1.6 dB, what is the RF Output Power of the Power Amplifier (PA) measured in W, dBW and dBm?

If the PA has a 36% efficiency, what is the required input power?

Exercise 7.3

(3 points) Find the beamwidth and diameter of a 6 dBi antenna, operating at 1.48 GHz, with an efficiency of 65%?

Exercise 7.4

(3 points) Consider a transmitter with an antenna efficiency of 0.65, a diameter of 4 m, and a frequency of 12 GHz.

1. Find the maximum gain of the antenna.

2. What is the beamwidth in degrees?

3. What is the pointing loss? Assume the pointing error of the antenna is 0.10°.

Exercise 7.5

(3 points) Consider the same transmitter with an input power of 20 W. If the transmitter has a polarization mismatch of 0.2 dB and a feeder loss of 0.5 dB, what is the EIRP of the transmitter?

Chapter 8

Propagation

Electromagnetic waves can move through any medium, even a vacuum! Unlike sound waves, which are shifts in a physical medium like air or water, electromagnetic waves are pure energy! But, this doesn't mean they travel through all mediums equally. An electromagnetic wave traveling through space may not lose any energy, but when traveling through ocean water, could lose signal strength. In this chapter, we'll discuss the **total propagation loss** as a signal travels from the transmitter to the receiver.

In link budgets, the total propagation loss can consist of a number of terms, including free space path loss, atmospheric loss, rain fade, and other weather-related impacts. These effects depend on the transmission frequency and can also depend on path length and geography of the terrestrial components of the link. The ITU has defined recommendations for computing degradation in a signal due to weather-related effects, but these can feel complex. In this text, we will only discuss free space path loss.

8.1 Free Space Path Loss (FSPL)

Free space path loss, FSPL or L_{FSPL}, represents the ratio of the received power to the transmitted power in a link between two isotropic antennas. As a signal traverses free space, the power decreases as a function of the distance squared. This is caused by the beam "spreading out" as it travels, distributing its power over larger and larger areas. If you're familiar with the concept of flux, this is the same reason that flux decreases with distance squared. A linear representation of the free space path loss is given as,

$$L_{\text{FSPL}} = \left(\frac{4\pi R}{\lambda}\right)^2 \text{ [unitless]} \tag{8.1}$$

where R is the distance between the two antennas and λ is the signal wavelength. It's important that R and λ have the same unit of distance, generally meters or kilometers. You may also see d used interchangeably with R to represent distance or path length.

Doggo Help

Doggo is attempting to swing its tail at supersonic speed to grab your attention. Bork! Keep in mind that you do not need to add the radius of the Earth to the R value used in the FSPL equations because this R represents the distance between the *ground station* and *satellite* antennas. It is admittedly confusing that we use the variable R in multiple places throughout this book.

The linear-space Equation 8.1 is unitless because the units in the numerator and denominator cancel out. We can equivalently calculate FSPL in decibels using:

$$L_{\text{FSPL}} = 20\log_{10}\left(\frac{4\pi R}{\lambda}\right) \text{ [dB]} \tag{8.2}$$

For reference, the FSPL of a GEO is typically on the order of 200 dB.

Example with Sat!

What's the FSPL to a GEO in L-band vs Ka-band? Your answer should be in units of decibels. How do the two values compare?

Solution:
Use Equation 8.2 to determine the FSPL for a GEO in L-band.

$$L_{\text{FSPL}} = 20 \log_{10} \left(\frac{4\pi R}{\lambda} \right)$$

$$= 20 \log_{10} \left(\frac{4\pi (35786 \text{ km})}{\left(\frac{3E5 \text{ km/s}}{1.5E9 \text{ Hz}} \right)} \right)$$

$$= 127.0 \text{ dB}$$

For Ka-band, we use the same procedure, but plug in a different frequency.

$$L_{\text{FSPL}} = 20 \log_{10} \left(\frac{4\pi R}{\lambda} \right)$$

$$= 20 \log_{10} \left(\frac{4\pi (35786 \text{ km})}{\left(\frac{3E5 \text{ km/s}}{20E9 \text{ Hz}} \right)} \right)$$

$$= 149.5 \text{ dB}$$

We see that as the frequency increases, so does the FSPL!

8.2 Path Length

Calculating the distance between a satellite and a ground station, or **path length**, is a tricky geometry problem. We can calculate the path length, d, between a satellite at altitude, h, and a ground station at a fixed location using the angle of elevation, α, at which the ground station sees the satellite above the horizon. You may have noticed that when we spoke about FSPL, we used the term R for path length and now we are using d.

Unfortunately, we cannot use a simple trigonometric relationship between elevation angle and altitude, as that would require the Earth to be flat. Instead, we will assume the Earth is spherical and utilize the Law of Sines and the Law of Cosines. Figure 8.1 illustrates the geometry involved and labels various variables that will be useful in our calculation.

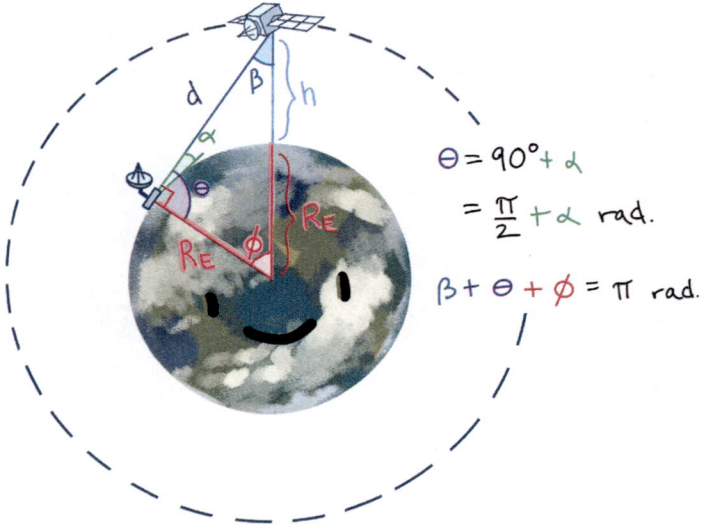

Figure 8.1: Path Length Geometry

For this derivation, we'll assume all distances use the same generic unit of length, and all angles are in radians. Applying the Law of Cosines with the center of the Earth yields Equation 8.3.

$$d^2 = R_{\mathrm{E}}^2 + (R_{\mathrm{E}} + h)^2 - 2R_{\mathrm{E}}(R_{\mathrm{E}} + h)\cos\phi \qquad (8.3)$$

where ϕ is the Earth's internal angle measured between the satellite and the ground station. Before solving for d, we need to find an expression for ϕ in terms of the angle α and other known quantities. To make this easier, let's define variables for the three internal angles (ϕ, θ, β) of our triangle, as shown in Figure 8.1.

$$\phi = \text{Earth internal angle} \tag{8.4}$$

$$\theta = \frac{\pi}{2} + \alpha \tag{8.5}$$

$$\beta = \pi - (\phi + \theta) \tag{8.6}$$

Our expression for θ sums to a right angle, $\pi/2$ radians or 90 degrees, below the horizon with α above the horizon, as shown in Figure 8.1, and β comes from the fact that the sum of the internal angles of any triangle is always π radians, or 180 degrees. With these internal angles, the Law of Sines yields:

$$\frac{\sin \phi}{d} = \frac{\sin \theta}{R_{\mathrm{E}} + h} = \frac{\sin \beta}{R_{\mathrm{E}}} \tag{8.7}$$

We can find an expression for β by rearranging the two rightmost terms of Equation 8.7:

$$\beta = \arcsin \left(\frac{R_{\mathrm{E}}}{R_{\mathrm{E}} + h} \sin \theta \right) \tag{8.8}$$

We now have an expression for β in terms of θ and an expression for θ in terms of α. This means we know both β and θ in terms of α alone, so we're ready to build our expression for ϕ. We'll start with the identity $\pi = \phi + \theta + \beta$ and rearrange terms to find:

$$\phi = \pi - (\theta + \beta) \tag{8.9}$$

Now we'll substitute in our expressions for θ and β to express ϕ solely in terms of α and given distances:

$$\phi = \frac{\pi}{2} - \alpha - \arcsin \left(\frac{R_{\mathrm{E}}}{R_{\mathrm{E}} + h} \cos \alpha \right) \tag{8.10}$$

Fantastic! The last step is to plug ϕ into our original Law of Cosines equation. This gives us our final path length equation:

$$d = \sqrt{R_{\mathrm{E}}^2 + (R_{\mathrm{E}} + h)^2 - 2R_{\mathrm{E}}(R_{\mathrm{E}} + h) \sin \left(\alpha + \arcsin \left(\frac{R_{\mathrm{E}}}{R_{\mathrm{E}} + h} \cos \alpha \right) \right)} \tag{8.11}$$

In this equation, R_E is the radius of Earth, h is the spacecraft altitude, and α is the satellite's elevation angle above the horizon *in radians*. Remember that the units of d must match the units of R_E and h, which must have the same unit of distance, e.g., meters or kilometers.

Doggo Help

Bork! Doggo dug up a math tip! There is another approach to calculating path length that involves utilizing the Law of Sines twice:

$$\frac{\sin \phi}{d} = \frac{\sin \beta}{R_E} \tag{8.12}$$

Next, we rearrange and substitute the angles we found earlier to obtain an equivalent expression for path length:

$$d = \frac{(R_E + h) \cos \left(\alpha + \arcsin \left(\frac{R_E}{R_E + h} \cos \alpha\right)\right)}{\cos \alpha} \tag{8.13}$$

Although the derivation for this solution requires fewer steps, it is undefined at $\alpha = \frac{\pi}{2}$ because $\cos(\frac{\pi}{2}) = 0$. You can use this equation if you prefer, just be aware that when the satellite is directly overhead this equation is undefined and the path length is h. Doggo quickly buries this math tip as Doggo does not like math as much as he likes belly rubs.

Example with Sat!

We mentioned that we can't use simple trigonometry to compute the path length because it assumes the Earth is flat. Quantify the difference between the "Earth is flat" trigonometric case and the "curved Earth" method derived from the Law of Cosines for a satellite at an altitude of 500 km and an elevation angle of $50°$, which is equivalent to $\frac{5\pi}{18}$.

Solution:
Using simplistic trigonometry, we can approximate path length as follows:

$$\sin \alpha = \frac{h}{d}$$

$$d = \frac{500 \text{ km}}{\sin\left(\frac{5\pi}{18}\right)} = 652.7 \text{ km}$$

Using the path length equation derived from the Law of Cosines:

$$d = \sqrt{\begin{aligned} &(6378 \text{ km})^2 + (6878 \text{ km})^2 - \\ &2(6378 \text{ km})(6878 \text{ km})* \\ &\sin\left(\frac{5\pi}{18} + \arcsin\left(\frac{6378}{6878}\cos\left(\frac{5\pi}{18}\right)\right)\right) \end{aligned}}$$

$$= 636.8 \text{ km}$$

That's a difference of nearly 16 km, which may seem small, but can increase with orbital altitude and become even more drastic!

8.3   Problem Set 6: Propagation

Problem	Topic	Points
1	GEO Path Length Computation	3
2	GEO Intersatellite Link FSPL Calculation	3
3	Two Approximations for FSPL at 6000 km	3
4	Propagation Loss	3
5	FSPL as a function of elevation angle	3
Total:		15

Exercise 8.1

(3 points) What is the path length from a GEO satellite communicating to an earth station with an elevation angle of 50°? Hint: Remember to convert 50° into radians.

Exercise 8.2

(3 points) Calculate the FSPL between inter-satellite links in GEO at 12 GHz. Be sure to include a sketch of your geometry. How does the FSPL change if your transmit frequency occurs at 1 GHz?

Exercise 8.3

(3 points) For a satellite orbiting at 6000 km, compute the path length using the simple trigonometric method, and the method derived using the Law of Cosine accounting for the curvature of the Earth. Assume an elevation angle of 50°. How do the two values compare?

Exercise 8.4

(**3 points**) What is the total propagation loss of a GEO satellite operating at 12 GHz with an elevation angle of 50°, and an atmospheric loss is 1.5 dB?

Exercise 8.5

(**3 points**) Write code, or use a tool like Excel, to parameterize your FSPL computation. Produce a figure of FSPL across elevation angles of 0 - 90 degrees for a satellite at the orbit of your choosing. Be sure to label your axis with proper units.

Chapter 9

Receivers (Rx)

So far, we have explored the first two of three primary components of a link: the transmitter and the propagation between the transmitter and receiver. Now, we will investigate the third major component: the receiver. Similar to transmitters, receivers consist of an antenna and modem. Unlike transmitters, receivers utilize a low-noise amplifier (LNA) directly after the antenna. The LNA increases the RF power of the signal that has just traveled hundreds or thousands of kilometers from the transmitter so that it can be accurately processed and received.

9.1 Received Power

It is critical that we design links with enough transmit power for the signal to successfully reach the receiver with a signal strength above the sensitivity level of the receiver. For this computation, we consider the received power, P_{Rx}, of a system.

To compute the P_{Rx}, or the strength of the signal in front of the Rx antenna we must remember that a surface of area, A, situated at distance, R, from the Tx antenna subtends a solid angle $\frac{A}{R^2}$. With this concept, we can calculate the strength of our transmitted signal using the equation:

$$P_{\mathrm{Rx}} = \left(\frac{P_{\mathrm{Tx}} G_{\mathrm{Tx}}}{4\pi} \right) \left(\frac{A}{R^2} \right) = \Phi A \; [\mathrm{W}] \qquad (9.1)$$

In this equation, P_{Tx} is the transmit power in Watts, G_{Tx} is the transmit antenna gain, and Φ is the power flux density (PFD). You may also see the surface area, A, described as the effective aperture area, A_{Reff}. It is important to note that this calculation is performed in linear space, so gain is expressed as a unitless power ratio and the output is measured in Watts.

Power flux density is an important concept in the world of satellite communications. The PFD is defined as the amount of power flowing through a unit area and is a commonly used metric for ensuring satellite systems won't cause harmful interference into terrestrial systems. In fact, Article 21 of the ITU's Radio Regulations is entirely devoted to PFD limits [16].

Figure 9.1 visualizes the distance, R, and effective surface area, A_{Reff}, of the receiver.

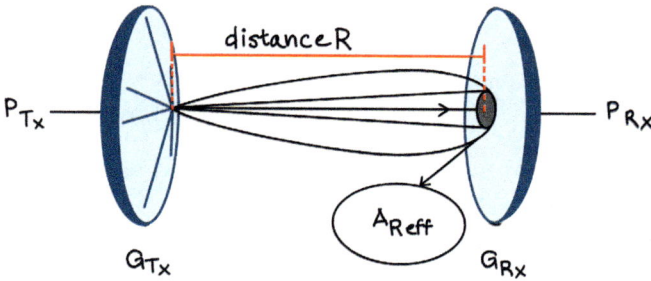

Figure 9.1: Effective Aperture of the Receiver

where the A_{Reff}, measured in m^2, is expressed as a function of the receiving gain, G_{Rx}:

$$A_{\text{Reff}} = \frac{G_{\text{Rx}}\lambda^2}{4\pi} \ [\text{m}^2]$$

(9.2)

Substituting this value for A into Equation 9.1 yields,

$$P_{\text{Rx}} = \left(\frac{P_{\text{Tx}}G_{\text{Tx}}}{4\pi}\right)\left(\frac{G_{\text{Rx}}\lambda^2}{4\pi R^2}\right)$$

(9.3)

Notice how the equation can be rewritten to the following form:

$$P_{\text{Rx}} = (P_{\text{Tx}}G_{\text{Tx}})\left(\frac{\lambda}{4\pi R}\right)^2 G_{\text{Rx}}$$

(9.4)

This second term might look familiar, and that is because it is! Recall Equation 8.1, which defines free space path loss, L_{FSPL}, as a unitless power ratio (i.e., *not* expressed in decibels).

$$P_{\text{Rx}} = (P_{\text{Tx}}G_{\text{Tx}})\left(\frac{1}{L_{\text{FSPL}}}\right)G_{\text{Rx}} \ [\text{W}]$$

(9.5)

The final form of the received power, including both transmit and receive losses, is known as the **Friis Equation**. The Friis Equation is one of the most important equations in this book! Note that the generic transmitter and receiver loss terms, L_{Tx} and L_{Rx}, respectively, will include losses, such as the feeder loss of the transmitter and receiver. The most common form of the Friis Equation is as follows:

$$P_{\text{Rx}} = \left(\frac{P_{\text{Tx}}G_{\text{Tx}}G_{\text{Rx}}}{L_{\text{Tx}}L_{\text{Rx}}} \right) \left(\frac{\lambda}{4\pi R} \right)^2 \text{ [W]} \qquad (9.6)$$

In decibel space, the Friis Equation becomes:

$$\begin{aligned} P_{\text{Rx,dB}} = P_{\text{Tx,dB}} + G_{\text{Tx,dB}} - L_{\text{Tx,dB}} \\ - L_{\text{FSPL,dB}} + G_{\text{Rx,dB}} - L_{\text{Rx,dB}} \end{aligned} \text{ [dBW]} \qquad (9.7)$$

where the "dB" subscript denotes the decibel expression of the corresponding term, e.g., $P_{\text{Tx,dB}} = 10\log_{10}(P_{\text{Tx}})$. Note that $-L_{\text{FSPL}} = 20\log_{10}(\frac{\lambda}{4\pi R})$, since flipping the fraction in the logarithmic term switches the sign of the free space path loss. Unfortunately, there is no standard notation for denoting linear or logarithmic terms in industry, so it will be critical for you to look at the context and units of a particular equation to determine whether the terms in question are linear or logarithmic. If the terms are linear, you will multiply and divide; if they're logarithmic, you will add and subtract.

Figure 9.2 provides a schematic to visualize each term and to enable us to compute the performance at the receiver, Rx.

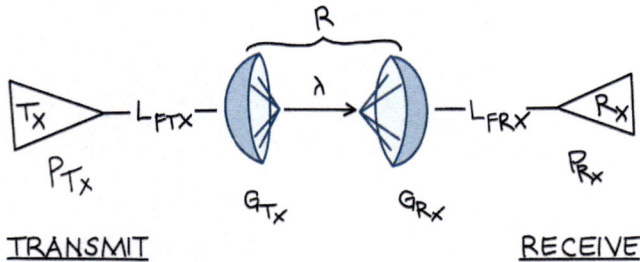

Figure 9.2: Communications System Schematic

Let's take a moment to practice some of the underlying concepts of the Friis Equation.

Consider the transmission from an ES to a GEO satellite at a Ka-band frequency of 27.5 GHz. By now, we hope that the altitude of a GEO is something you remember!

1. What is the receive gain, G_{Rx}, of a system with a 3 dB beamwidth of 2° and an efficiency of 0.55?

2. If the EIRP of the Tx is 60 dBW, what is the received power at the GEO satellite?

Solution:

1. To calculate G_{Rx}, we must first calculate the antenna diameter, d. We can do this using the beamwidth equation shown in Equation 7.8.

$$\theta_{3\text{dB}} = 70(\frac{c}{fd})$$

$$2° = 70\left(\frac{3E8 \text{ m/s}}{(27.5E9 \text{ Hz})d}\right)$$

$$d = 0.38 \text{ m}$$

We can then plug the diameter into Equation 7.4 to find G_{Rx}:

$$G_{Rx} = 10 \log_{10} \eta \left(\frac{\pi d}{\lambda}\right)^2$$

$$= 10 \log_{10} 0.55 \left(\frac{\pi(0.38 \text{ m})}{\frac{3E8 \text{ m/s}}{27.5E9 \text{ Hz}}}\right)^2$$

$$= 38.2 \text{ dBi}$$

Example with Sat!

2. Using the answer to the first part of this problem, and given that EIRP = 60 dBW, we can calculate the received power with Equation 9.7, assuming a loss of 0 dB. Keep in mind that EIRP is the sum of our transmit power and transmit gain (the first two terms of the Friis Equation).

$$P_{\text{Rx}} = EIRP + G_{\text{Rx}} - L_{\text{FSPL}}$$

$$P_{\text{Rx}} = 60 \text{ dBW} + G_{\text{Rx}} - 20\log_{10}\left(\frac{4\pi R}{\lambda}\right)$$

$$= 60 \text{ dBW} + 38.2 \text{ dBi} -$$

$$20\log_{10}\left(\frac{4\pi(35786E3 \text{ m})}{\frac{3E8 \text{ m/s}}{27.5E9 \text{ Hz}}}\right)$$

$$= 60 \text{ dBW} + 38.2 \text{ dBi} - 212.3 \text{ dB}$$

$$= -114.1 \text{ dBW}$$

9.2   Receiver Performance (G/T)

We previously saw that EIRP is used to compute the overall performance of a transmitter. Conversely, for receive antennas, we use a metric called G/T to compute the overall receiver performance. Like "E-I-R-P" or "EIRP", G/T is referred to in a variety of ways, "G over T", "G on T", or "gain over temperature." Numerous factors can be incorporated into a G/T calculation, making this metric the most common source of discrepancy in a link budget. In principle, G/T is calculated as,

$$G/T = 10\log_{10}(G_{\text{Rx}}) - 10\log_{10}(T) - \Sigma L_{\text{Rx}} \text{ [dB/K]} \qquad (9.8)$$

where G_{Rx} is the receiver antenna gain, T is a "noise temperature" which we'll cover in detail later in this chapter, and ΣL_{Rx} represents whatever set of receiver-side losses make sense to include. The units of G/T are reported in dB/K.

Doggo Help

Woof! Doggo wonders why everything has so many units, and why we don't measure everything in bones instead? By convention, G/T has the strange unit of dB/K. This arises from the fact that receiver antenna gain is given in dBi, the losses in dBW, and the noise temperature in dBK. We recall that dBi references an isotropic antenna, dBW references Watts and dBK references Kelvins, and that subtraction in logarithmic space is division in linear space. Therefore, when we compute $G - T - L$, we're creating a ratio where the numerator and denominator have different reference units. It's cumbersome to report the units as dBi/dBW/dBK or dB(i/W/K), so by convention, we report the units as dB/K. It's a bit of a mess, but at least there's a standard!

To compute this parameter, we must go a step beyond the power hitting the receive antenna, and consider the LNA and feeder. We have already learned how to compute the gain in G/T and that the denominator, T, stands for systems noise temperature. In the next section, we will define T to be T_{sys}. For now, we will assume a value for our temperature and practice computing G/T.

Example with Sat!

Given a receiver operating at 10 GHz. Assume the antenna diameter is 1.5 m with an aperture efficiency of 0.65 and a system noise temperature of 500 K.

1. What is the maximum receive gain, G_{Rx}, of the system?

2. What is the G/T, measured in dB/K?

Solution:

1. We use Equation 7.6 to find the max gain:

$$G_{\max} = 10 \log_{10} \eta \left(\frac{\pi d}{\lambda} \right)^2$$

$$= 10 \log_{10} 0.65 \left(\frac{\pi(1.5 \text{ m})}{\frac{3E8 \text{ m/s}}{10E9 \text{ Hz}}} \right)^2$$

$$= 42.1 \text{ dBi}$$

2. To find G/T, we use Equation 9.8:

$$G/T = 42.1 \text{ dBi} - 10 \log_{10}(500 \text{ K})$$

$$= 15.1 \text{ dB/K}$$

9.3   System Noise Temperature, T_{sys}

When we are designing our system, we have to compute the system noise temperature. Space is filled with low-frequency background radiation, e.g., infrared and radio-frequency radiation from stars, planets, and other interstellar objects. In addition to measuring a desired signal, a receiving antenna picks up this background "noise" as well. **Noise temperature** quantifies the amount of power that a receiving antenna absorbs from background sources of electromagnetic radiation and is measured in Kelvin. Unfortunately, this quantity can be computed at multiple locations in the chain of receiver hardware, further contributing to potential confusion.

Before we study the system noise temperature equations, let's take a look at a schematic of a receiver and two possible locations for computing our system noise temperature. Figure 9.3 offers two locations where the temperature may be calculated, T_1, which is located directly behind the antenna and before the feeder, or T_2, which is located after the feeder. It is important to note that the temperature calculations are often performed in linear space!

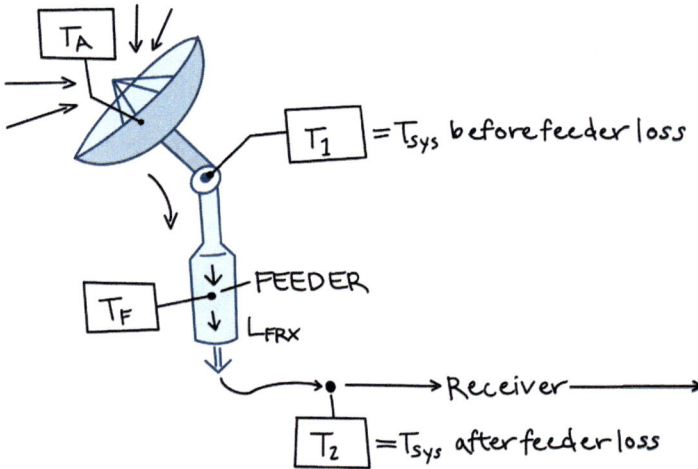

Figure 9.3: Two Locations for Calculating System Temperature

T_1 shows a system noise temperature computed right at the antenna output, before the feeder loss. The linear equation for T_{sys}, measured in Kelvin is:

$$T_1 = T_A + T_F(L_{\text{FRX}} - 1) + \frac{T_{\text{eff}}}{L_{\text{FRX}}} \ [\text{K}] \tag{9.9}$$

where T_A, T_F, and L_{FRX} represent antenna noise temperature, feeder noise temperature, and receiver feeder loss, respectively. We'll explain them in more detail in a moment. If you instead compute the system noise temperature at the input to the receiver, after the feeder loss, then the system noise temperature, T_2, becomes:

$$T_2 = \frac{T_A}{L_{\text{FRX}}} + T_F\left(1 - \frac{1}{L_{\text{FRX}}}\right) + T_{\text{eff}} \ [\text{K}] \tag{9.10}$$

This is equivalent to $T_2 = \frac{T_1}{L_{FRX}}$, except the T_{eff} term is not divided by L_{FRX}. When calculating your final link, either T_1 or T_2 may be used for T_{sys}. In this book, we will use the temperature at the receiver input, $T_2 = T_{\text{sys}}$.

9.4   Antenna Noise Temperature, T_A

An antenna picks up noise from any radiating body within the radiation pattern of the antenna. Noise is therefore based on the direction in which the receiver antenna is pointing, its radiation pattern, and the state of the surrounding environment. In our system, the antenna is characterized by the antenna noise temperature, T_A. For an antenna on the ground with a field of view of the sky, T_A consists of a sky noise temperature, T_{SKY}, and a ground noise temperature, T_{GROUND}, which together capture background noise in the field of view of the antenna. Figure 9.4 shows an illustration of this scenario and T_A computation.

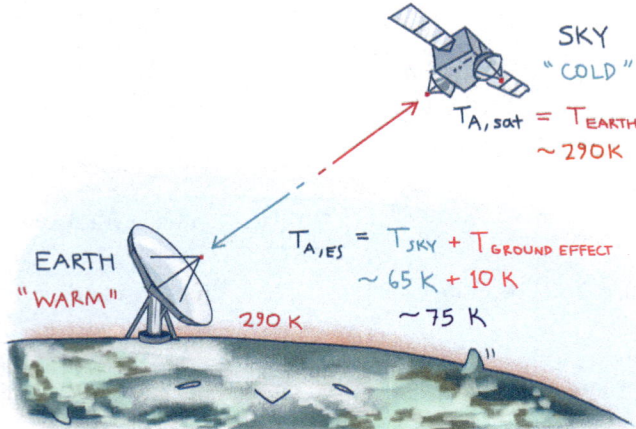

Figure 9.4: Earth vs. Space Station Antenna Temperature, T_A

If the receiving antenna is on a satellite, it is likely pointed down at Earth, which is a relatively "warm" radiating body. If the antenna is an Earth Station (ES), it is likely looking at the sky, which is relatively "cold", but must include a contribution from the "warm" ground temperature, T_{GROUND}, which is generally 10 K for elevation angles of 10 to 90 degrees. In summary, T_A is often assumed to be 290 K for a receiver antenna on a satellite pointed at the Earth and 75 K for an antenna on the ground pointed at the sky. To sanity check your work, remember that the Earth is far noisier than looking up at a blank sky!

Doggo Help

Confused by all the terms and numbers we just threw around? Here's a summary of the important ideas:

- T_{A} is the *antenna noise temperature* that arises from background radiation.

- T_{A} is higher when the antenna is pointed at a "warm" radiating body like the Earth as opposed to the sky.

- For a satellite receiver pointed at the warm Earth, it is safe to assume a value of $T_{\mathrm{A,sat}} = 290$ K.

- For an ES receiver pointed at the cold sky, it is safe to assume $T_{\mathrm{A,ES}} = 75$ K.

9.5 Effective Noise Temperature, T_{eff}

Because T_{sys} arises from the various components of the receiver system, it includes contributions from the amplifier's noise temperature, T_{amp}, which can alternatively be named the effective noise temperature, T_{eff}. This is in addition to the antenna noise temperature, T_{A}. T_{eff} is the thermodynamic temperature of a resistance which, placed at the input of the element assumed to be noise free, establishes the same noise power at the output of the element as the actual element without noise. An expression for T_{eff} is given in Equation 9.11.

$$T_{\mathrm{amp}} = T_{\mathrm{eff}} = T_0(10^{\frac{\mathrm{NF}}{10}} - 1) \ [\mathrm{K}] \tag{9.11}$$

where NF is the LNA "noise figure" and T_0 is a standard reference noise temperature, usually 290 K. The *noise figure* of an LNA is the ratio of total available noise power at the output of the actual device to the noise output of an "ideal" receiver with the same overall gain and bandwidth when the receivers are connected to matched sources at the standard reference noise temperature. If a value for the LNA NF is not known, an approximation in the range 0.1 - 2 dB is appropriate.

We now have all of the terms required to compute the system temperature, T_{sys}, and the overall performance of the receiver.

9.6   Problem Set 7: Receivers

Problem	Topic	Points
1	Receive Power at OneWeb Satellite	3
2	Basic Gain over Temperature Computation	3
3	System Noise Temperature	3
4	Antenna Noise Temperature	1
Total:		10

Exercise 9.1

(**3 points**) Compute the received power at a OneWeb satellite from a Ku-band UT that transmits at 14.0 GHz with a P_{Tx} of 2 Watts and a G_{Tx} of 34 dBi. Assume the satellite G_{Rx} is 25 dBi.

Exercise 9.2

(**3 points**) If the satellite's G_{Rx} is 37.8 dBi with 1.6 dB of loss, what is the G/T, if the system noise temperature is 500 K?

Exercise 9.3

(**3 points**) Compute T_{sys}, given an antenna temperature of T_{A} = 75 K, an LNA NF = 2 dB, and a feeder loss of 1 dB.

Exercise 9.4

(**1 point**) Given the scenario in the previous problem, what would you assume T_{A} to be if your transmission was in the opposite direction, meaning an uplink instead of a downlink? Why?

Chapter 10

Link Performance

Hooray! We have now covered all three aspects of a link: the transmitter, the receiver, and the medium in between. In this chapter, we will assess the overall link performance, generally characterized in terms of signal-to-noise ratio, or SNR.

10.1 Signal to Noise Ratio (SNR)

In addition to the **signal-to-noise ratio**, SNR, we can also measure performance using the carrier-to-noise power ratio, C/N, or the C/N_0, which means the carrier-to-noise power ratio in one Hertz of spectrum instead of the entire channel bandwidth. It is also possible that you may see C/N+I, which accounts for interference, or other types of metrics like Eb/No and Es/No. More on these soon!

Data is generally encoded into a signal by modulating a wave of some base frequency. Amplitude-modulated (AM) signals encode data by varying the amplitude of this base wave, known as the carrier, while frequency-modulated (FM) signals vary the frequency slightly. When quantifying the performance of a link, the strength of an incoming carrier wave, C, can be calculated as:

$$C = \text{EIRP} - L_{\text{FSPL}} + G_{\text{Rx}} \text{ [dB]} \tag{10.1}$$

Each of the terms in the above equation is expressed in decibels.

Noise, N, consists of unwanted external radiation that can impact the carrier power. By adding unwelcome fluctuations to the received signal, noise reduces the ability of the receiver to correctly decode the information stored in the carrier. Noise can come from natural sources within the vicinity of the antenna reception area as well as imperfections in the hardware components of the receiver. Signals from other transmitters that we don't intend to receive also manifest in the system as noise, but are generally accounted for as *interference*.

To calculate noise power we use the equation $N = kTB$, where k is Boltzmann's constant ($k = 1.380649E{-}23$ J/K), T is noise temperature, and B is bandwidth. Expressed in decibel space, noise can be calculated as:

$$N = 10\log_{10}(k) + T_{\text{sys}} + 10\log_{10}(B) \text{ [dB]} \tag{10.2}$$

where $10\log_{10}(k) = -228.6$ dBW/K/Hz and T_{sys} is the receiver system noise temperature that we calculated with Equation 9.10.

The SNR metric, C/N, can be computed directly from Equations 10.1 and 10.2 as

$$\text{C/N} = C - N \text{ [dB]} \tag{10.3}$$

or can alternatively be calculated as:

$$\text{C/N} = \text{EIRP} - L_{\text{FSPL}} - \text{G/T} + 228.6 - 10\log_{10}(B) \text{ [dB]} \tag{10.4}$$

In this form, additional carrier signal losses such as atmospheric losses, L_{atm}, may be integrated into this equation by subtracting them alongside L_{FSPL}. Pay attention to the reversed signs for Boltzmann's constant and bandwidth, as these are due to the subtraction of N in the calculation. Additionally, it's important to understand that G/T, which serves as a measure of performance for the receiver, does not represent a simple division of receiver gain by system noise temperature.

> **Doggo Help**
>
> Bork! The "carrier-to-noise ratio" is the ratio of the strength of the carrier to the strength of the noise, so it makes sense to represent it with the label "C/N". However, C/N is expressed as a decibel, for which the mathematical operation is $C_{\text{dB}} - N_{\text{dB}}$ (the subscripts meaning both are in decibels). This leads to the confusing statement that C/N $= C_{\text{dB}} - N_{\text{dB}}$, which reads as "$C$ divided by N" in linear space equals "C minus N" in decibel space.
>
> The label "C/N" clearly communicates the concept of a ratio between carrier and noise, so it is convenient to label that metric as C/N. Nevertheless, we calculate and represent this metric in decibel space, so we must recognize that C/N represents a concept and not a mathematical operation. The same goes for the label G/T, which represents the gain-to-temperature ratio of a receiver system. Doggo is gaining motivation to create a bone-based unit system. It may not be better, but it will be more delicious.

In the real world, systems typically use a unique bandwidth, which is tied to data rate and frequency, so C/N cannot be used to compare one company's system to another directly. To allow for direct comparisons, we can "normalize" C/N by bandwidth. C/N_0 is a metric that effectively takes all the C/N across a system's bandwidth and crams it into 1 Hz and has units of dB-Hz. This is not a physical quantity but a metric that allows the C/N's of various systems to be directly compared. C/N_0 may be calculated as:

$$\text{C/N}_0 = \text{EIRP} - L_{\text{FSPL}} - G/T + k \text{ [dB-Hz]} \tag{10.5}$$

which is exactly the same as Equation 10.4 without the bandwidth term.

Doggo Help

Doggo brings you a math tip that fell from space. Woof! Confused by how the subtracted bandwidth term relates to normalization? It's actually easier to think about this in linear space:

Equation 10.4 calculates C/N in decibel space, so it's mathematically equivalent to C/N= $\frac{\text{EIRP}}{L_{\text{FSPL}}*G/T*kB}$ in linear space (remembering again that G/T is a *quantity*, not a fraction). We can interpret this as saying that C/N is equal to the quantity $\frac{\text{EIRP}}{L_{\text{FSPL}}*G/T*k}$, divided across some bandwidth, B. We call this fractional quantity C/N$_0$, and because it does not rely on bandwidth, we can directly compare it across systems. To calculate C/N$_0$, we multiply C/N by B in linear space, equivalent to C/N+$10\log_{10}(B)$ in decibel space. This decibel addition cancels out the subtracted bandwidth term in Equation 10.4, leaving us with Equation 10.5.

Phew! Decibels can be weird to wrap our heads around, but they're pretty elegant with a bit of interpretation!

Another metric commonly used to compare different systems is the "energy per bit to noise spectral density" ratio, or E_b/N_0. This metric is pronounced "ebb-no", and is another normalized signal-to-noise ratio metric, commonly referred to as "SNR per bit." This metric can be calculated in decibel space in the following manner:

$$E_b/N_0 = \text{EIRP} - L_{\text{FSPL}} - \text{G/T} + 228.6 - 10\log_{10}(R) \text{ [dB]} \qquad (10.6)$$

where R is the data rate in bits per second, and is equal to:

$$E_b/N_0 = \text{C/N}_0 - 10\log_{10}(R) \text{ [dB]} \qquad (10.7)$$

From E_b/N_0, one can compute C/N as:

$$\text{C/N} = E_b/N_0 + 10\log_{10}(R) - 10\log_{10}(B) \text{ [dB]} \qquad (10.8)$$

Doggo Help

Bork! Not convinced by the relationship between C/N and E_b/N_0? Let's think in **linear space** for a moment so that we can more easily consider the units here:

$$C/N = \frac{E_b}{N_0} \times \frac{R}{B} \tag{10.9}$$

E_b is measured in units of energy per bit, and R is measured in bits per second. Multiplying the two,

$$E_b \left[\frac{J}{\cancel{bit}}\right] \times R \left[\frac{\cancel{bits}}{s}\right] \tag{10.10}$$

results in energy per second, or Watts. The product of $E_b \times R$, yields a power! This quantity represents the total power contained in the carrier wave, C. We can also recall that N_0 is the noise spectral density, which has units of power per Hz, and B is bandwidth, measured in Hz.

$$N_0 \left[\frac{W}{\cancel{Hz}}\right] \times B \, [\cancel{Hz}] \tag{10.11}$$

results in Watts, another power! This quantity represents the total power contained in the background noise, N. Now let's revisit the ratio between the two:

$$\frac{E_b \times R \, [W]}{N_0 \times B \, [W]} = \frac{P_{\text{carrier}} \, [\cancel{W}]}{P_{\text{noise}} \, [\cancel{W}]} = C/N \tag{10.12}$$

C/N correctly comes out as dimensionless! Don't forget to convert back into decibels!

Example with Sat!

What is the EIRP, C/N$_0$, and C/N for a 1.625 MHz bandwidth link to a GEO satellite operating at 14.25 GHz given:

- $G_{\text{Tx}} = 28.5$ dBi

- $P_{\text{Tx}} = 1$ W

- $L_{\text{Tx}} = 1.4$ dB

- G/T = -4.95 dB/K

- $L_{\text{atm}} = 1.5$ dB

Solution:
First, we must calculate EIRP:

$$EIRP = G_{\text{Tx}} + P_{\text{Tx}} - L_{\text{Tx}}$$
$$= 28.5 \text{ dBi} + 10\log_{10}(1 \text{ W}) - 1.4 \text{ dB}$$
$$= 27.1 \text{ dBW}$$

To calculate C/N, we need to simplify L_{FSPL} using Equation 8.2.

$$L_{\text{FSPL}} = 20\log_{10}\left(\frac{4\pi R}{\lambda}\right)$$
$$= 20\log_{10}\left(\frac{4\pi(35786 \text{ km})}{\frac{3E5 \text{ km/s}}{14.25E9 \text{ Hz}}}\right)$$
$$= 206.60 \text{ dB}$$

We can now substitute this value when computing C/N.

$$\text{C/N} = \text{EIRP} - L_{\text{FSPL}} - L_{\text{atm}} + \frac{G_{\text{Rx}}}{T_{\text{sys}}} - k - 10\log_{10}(B)$$
$$= 27.1 \text{ dBW} - 206.6 \text{ dB} - 1.5 \text{ dB} - 4.95 \text{ dB/K} -$$
$$(-228.6 \text{ dB}) - 10\log_{10}(1.625E6 \text{ Hz})$$
$$= -19.45 \text{ dB}$$

Example with Sat!

Lastly, we calculate C/N_0:

$$C/N_0 = C/N + 10\log_{10}(B)$$
$$= -19.45 \text{ dB} + 10\log_{10}(1.625E6 \text{ Hz})$$
$$= 42.66 \text{ dB}$$

10.2 Eb/No and Es/No

Earlier in the chapter, we introduced the concept of energy per bit to noise spectral density, Eb/No. Since the signal power is measured in Watts and the data rate in bits per second (bps), the unit for Eb (energy per bit) is Joules (J), which is equivalent to Watt-seconds. The noise spectral density, No, indicates the noise power in 1 Hz of bandwidth, and is also measured in Watts per Hertz (W/Hz) or Joules. Thus, the ratio Eb/No is dimensionless, meaning it has no units. Depending on the specifications of a given component, some might prefer using the ratio of energy per symbol to noise spectral density, Es/No, as a measurement for system performance. For example, many GEO modem providers specify modem performance using Es/No, so individuals who work in this space rarely use Eb/No. The only difference here is the amount of energy in a symbol versus a bit.

Converting an analog signal to a digital signal and vice versa is known as modulation. The word "modem" refers to a device that is responsible for *mo*dulation and *dem*odulation, hence the word *"mo-dem"*. Binary phase shift keying (BPSK) is an example of a modulation technique that maps each bit (1 or 0) to two different phase states. Figure 10.1 provides a schematic of a BPSK modulated and binary signal.

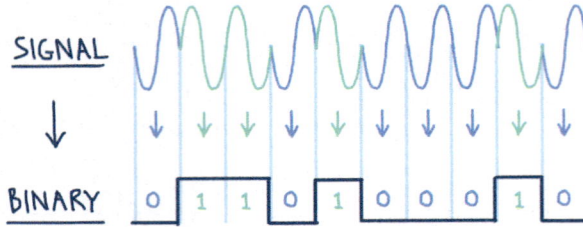

Figure 10.1: BPSK Raw and Modulated Signal

In addition to BPSK, several other types of modulation schemes exist: QPSK, offset QPSK or OQPSK, 8PSK, 16PSK, 16QAM, 16APSK, 32APSK, 64QAM. Constellation diagrams pictorially represent a signal modulated by a digital modulation scheme. The constellation diagrams for a few of the aforementioned types of modulation schemes are shown in Figure 10.2. Quadrature Phase Shift Keying (QPSK) maps 2 bits to each symbol such that the modulated signal can take 1 of 4 phase shapes.

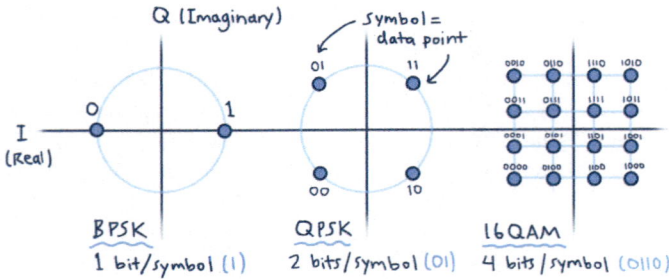

Figure 10.2: Types of Modulation Schemes

Figure 10.2 also lists the number of bits per symbol for each modulation scheme. This is useful when converting from Eb/No to Es/No. Since there are generally multiple bits per symbol, the energy in a symbol should be larger than that of a single bit, so to convert from Eb/No to Es/No we add $10\log_{10}$ (bits/symbol).

Example with Sat!

How does the Eb/No compare to the Es/No of a link with BPSK, QPSK, and 16QAM modulation?

Solution:
For BPSK, each symbol represents one bit, so Es = Eb. In this case, the Eb/No equals the Es/No.

For QPSK, each symbol represents 2 bits, so Es = 2*Eb. This means that Es/No is twice the value of Eb/No.

For 16QAM, each symbol represents 4 bits, so Es = 4*Eb. This means that Es/No is four times the value of Eb/No.

10.3 Spectral Efficiency and Data Rate

The spectral efficiency is the data rate per unit of spectrum, measured in bits/second/Hz (or bps/Hz). It and bit error rate (BER) are important performance metrics of digital modulation schemes. Systems often target BERs on the order of 10E-05 or 10E-07. The theoretical curve in Figure 10.3 follows an error function and enables us to map values of Eb/No to BER.

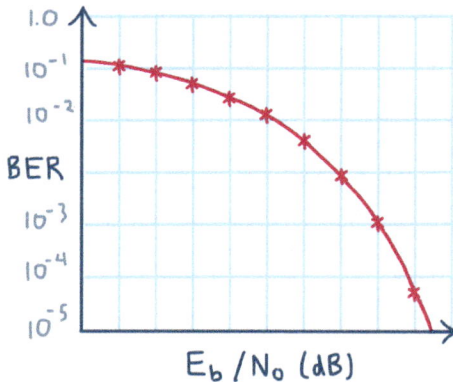

Figure 10.3: Mapped Eb/No to BER

If you wanted to determine the required Eb/No for a system with a BER of $10E-03$, you could use the curve provided in Figure 10.3, and backout the value of Eb/No that relates to a BER of $10E-03$.

Sometimes when constructing link budgets, we are given a required data rate, bit error rate, and margin. Other times, we are given the inputs of the system like antenna performance, orbital altitude, and modem performance and are asked to compute the achievable data rate or spectral efficiency of the system.

The maximum data rate I_{max} in a channel, or the theoretical channel capacity, can be found using Shannon's Limit [29]:

$$I_{max} = B \log_2 (1 + C/N) \text{ [bps]} \tag{10.13}$$

This is a "theoretical limit" of a channel, so as a sanity check, it can be helpful to compute the data rate in your system and compare it against Shannon's limit. If your calculated data rate is greater than Shannon's limit, you might have a bug, or your system might be over-specified - turn down your power!

Example with Sat!

Show that the maximum channel capacity of a 1625 kHz channel with a C/N of -20 dB is ~ 23 kbps. Remember to input C/N in a linear scale and not in decibels.

Solution:

$$I_{max} = B \log_2 (1 + C/N)$$
$$= (1.625E3 \text{ Hz}) \times \log_2 \left(1 + 10^{\frac{-20 \text{ dB}}{10}}\right)$$
$$= 23327 \text{ bps}$$
$$\approx 23 \text{ kbps}$$

10.4 Problem Set 8: Link Performance

Problem	Topic	Points
1	Boltzmann's Constant	1
2	C/N and C/No	3
3	System Performance of LEO Smallsat	3
4	Impact of LNA NF on C/N	3
5	System Performance of GEO	3
6	End to End Link Budget	3
Total:		16

Exercise 10.1

(**1 point**) What is the value of Boltzmann's constant, k, in dB? Make sure to use the proper units prior to converting.

Exercise 10.2

(**3 points**) Given a 250 MHz channeled system has a carrier strength, C of -94 dBW, and a noise power of -108 dB. Compute the C/N and C/N_0.

Exercise 10.3

(**3 points**) What is the EIRP, C/N_0, and C/N for a 250 MHz link to a smallsat at 550 km with a transmit frequency of 2.025 GHz given:

- $P_{Tx} = 0.75$ W

- $G_{Tx} = 16.5$ dBi

- $L_{\mathrm{Tx}} = 2$ dB

- $\mathrm{G/T} = -2$ dB/K

- $L_{\mathrm{atm}} = 1.5$ dB

Exercise 10.4

(**3 points**) What impact do the following parameters have on C/N?

- A 1 dB increase or decrease in the LNA NF

- Doubling the Rx antenna diameter

- An increase in elevation angle from 45 degrees to 90 degrees

Exercise 10.5

(**3 points**) What is the received power, or C, for a 4 GHz GEO link (use an altitude of 35786 km), with a 100 MHz bandwidth, given a transmit power of 100 W, an elevation angle of 60 degrees, an antenna transmit gain of 18 dBi, and a parabolic receive dish of 3 m? What is the G/T if the T_{sys} is 290 K and the antenna efficiency is 0.65? If the Eb/No required is 10 dB, what is the achievable data rate?

Exercise 10.6

(**3 points**) What is the receive power at a 1.5 meter satellite antenna, operating at 40 GHz, if the transmit earth station is 2.4 meter in diameter with a 200 W transmit power. Assume an orbital altitude of 1000 km and an elevation angle of 65 degrees. Both antennas have an aperture efficiency of 0.65.

Chapter 11

Conclusion

Wow, what a journey! In this textbook, you learned about the basics of orbital mechanics and radio communications, two vital concepts for operating satellite systems. Let us look back and summarize the concepts we have seen:

- **Chapter 2: Orbit Types and Elements** covered orbit types (LEO , MEO , GEO, HEO, SSO) and orbital parameters. Remember: a is for the semi-major axis, e is for eccentricity, i is for inclination, Ω is for RAAN (Right Ascension of Ascending Node), ω is for the argument of perigee, and ν is for true anomaly.

- **Chapter 3: Orbital Mechanics: 2D** gave us a way to understand orbits through a physics lens using Kepler's Laws of Planetary Motion and Newton's Laws, and how to quantify the orbital period and the orbit equation.

- **Chapter 4: Hohmann Transfers** revealed a particularly important method for maneuvering satellites from one circular orbit to another.

- **Chapter 5: Space Sustainability** provided an overview of the space environment, space debris, post-mission disposal, and end-of-life maneuvers.

- **Chapter 6: Link Design & Analysis** opened our world to radio communications, summarizing how frequency in the spectrum is allocated and providing an introduction to decibels.

- **Chapter 7: Transmitters** gave us an overview of how signals, created from data, are propagated into space. Important parameters of transmitters include signal power, signal gain, pointing loss,

141

polarization, and the summary characteristic of all transmitters: EIRP.

- **Chapter 8: Propagation** explored signal behavior as they travel through space. In this chapter, we also learned about free space path loss (FSPL) and methods for calculating path length.

- **Chapter 9: Receivers** considered the other side of the link design process: receiving and decoding the signal. We also evaluated receiver performance through receiver power, gain (G/T), antenna noise temperature (T_a), and effective noise temperature (T_{eff}).

- **Chapter 10: Link Performance** wrapped up our learning journey with a means for calculating how well a link design works from the transmitter, through space, and to the receiver. We covered the signal-to-noise ratio (SNR) as a means of calculating how strong our signal is after going through a communications link. We also explored various parameters that should be considered when designing a link budget, such as the energy per bit to noise spectral density (Eb/No), spectral efficiency, and data rate.

If something in these summaries sticks out to you as unfamiliar, feel free to review that chapter and refresh yourself. Otherwise, congratulations! You have successfully mastered the fundamentals of satellite communications systems!

HOORAY!
SHOOT FOR THE STARS

Part III

Appendix

Appendix A

Equation Sheet

A.1 Chapter 2: Orbit Types and Elements

- Semi-major axis: $a = \frac{r_a + r_p}{2}$

- Eccentricity: $e = \frac{r_a - r_p}{r_a + r_p}$

A.2 Chapter 3: Orbital Mechanics 2D

- Newton's Law of Gravitation: $F = \frac{GMm}{r^2}$

- Orbital Period: $T = \frac{2\pi}{\sqrt{\mu}} a^{\frac{3}{2}}$

- Specific angular momentum: $\vec{h} = \frac{\vec{L}}{m} = \vec{r} \times \vec{v}$

- Energy conservation: $\epsilon = -\frac{\mu}{2a}$

- Vis-viva: $\epsilon = \frac{v^2}{2} - \frac{\mu}{r}$

- Orbit equation: $r(\theta) = \frac{h^2}{\mu} \frac{1}{(1 + e\cos\theta)}$

- Orbit equation (with semi-major axis): $r(\theta) = a \frac{1 - e^2}{(1 + e\cos\theta)}$

A.3 Chapter 4: Hohmann Transfers

- Change in velocity (first burn): $\Delta v_1 = v_{\text{elliptical}} - v_{\text{circular}_{\text{initial}}} = \sqrt{\frac{2\mu * r_f}{r_0(r_0 + r_f)}} - \sqrt{\frac{\mu}{r_0}}$

- Change in velocity (second burn): $\Delta v_2 = v_{\text{circular}_{\text{final}}} - v_{\text{elliptical}} = \sqrt{\frac{\mu}{r_f}} - \sqrt{2\mu \frac{r_0}{r_f(r_f+r_0)}}$

A.4 Chapter 5: Link Design

- Converting a value a non-logarithmic power ratio (linear space) into dB space:
$$X \text{ dB} = 10 \log_{10}\left(\frac{P_{\text{measured}}}{P_{\text{reference}}}\right)$$

- Converting a value from dB space into linear space: Power Ratio $= 10^{\left(\frac{X \text{ dB}}{10}\right)}$

- Addition in log space: $\log(A) + \log(B) + \log(C) = \log(A * B * C)$

- Subtraction in log space: $\log(A) - \log(B) = \log(\frac{A}{B})$

A.5 Chapter 6: Transmitter

- Gain: $G_{\text{max}} = \left(\frac{4\pi}{\lambda^2}\right) A_{\text{eff}}$

- Gain, in terms of efficiency: $G_{\text{max}} = \eta(\frac{\pi d}{\lambda})^2$

- 3 dB beamwidth (in degrees): $\theta_{3\text{dB}} = 70\left(\frac{\lambda}{D}\right) = 70\left(\frac{c}{fd}\right)$

- Pointing loss: $L_{\text{Pointing}} = 12\left(\frac{\theta_{\text{Pointing}}}{\theta_{3\text{dB}}}\right)^2$

- Polarization mismatch: $L_{\text{Polarization}} = -20\log\cos\psi$

- Effective Isotropic Radiated Power: $EIRP = G_{\text{Tx}} + P_{\text{Tx}} - L_{\text{Pointing}} - L_{\text{Polarization}} - L_{\text{Feeder}}$

A.6 Chapter 7: Propagation

- Free Space Path Loss (in linear space): $L_{\text{FSPL}} = \left(\frac{4\pi R}{\lambda}\right)^2$

- Free Space Path Loss (in dB): $L_{\text{FSPL}} = 20\log\left(\frac{4\pi R}{\lambda}\right)$

- Path length (using Law of Cosines): $d^2 = R_{\text{E}}^2 + (R_{\text{E}} + h)^2 - 2R_{\text{E}}(R_{\text{E}} + h)\cos\phi$

- Path length (using Law of Sines): $\frac{\sin\phi}{d} = \frac{\sin\theta}{R_{\text{E}}+h} = \frac{\sin\beta}{R_{\text{E}}}$

- Angle β (refer to Figure 8.1): $\beta = \arcsin(\frac{R_E}{R_E+h}\sin\theta)$

- Earth Internal Angle ϕ: $\phi = \pi - (\theta + \beta) = \frac{\pi}{2} - \alpha - \arcsin(\frac{R_E}{R_E+h}\cos\alpha)$

- Path length (plugging ϕ into original Law of Cosines equation):
$d = \sqrt{R_E^2 + (R_E + h)^2 - 2R_E(R_E + h)\sin(\alpha + \arcsin(\frac{R_E}{R_E+h}\cos\alpha))}$

- Path length (utilizing Law of Sines twice): $\frac{\sin\phi}{d} = \frac{\sin\beta}{R_E}$

- Path length (rearranging and substituting angles):
$d = \frac{(R_E+h)\cos((\alpha+\arcsin(\frac{R_E}{R_E+h}\cos\alpha))}{\cos\alpha}$

A.7 Chapter 8: Receivers

- Received power: $P_R = \left(\frac{P_T G_T}{4\pi}\right)\left(\frac{A}{R^2}\right) = \Phi A = \frac{P_T G_T}{4\pi R^2}A_{\text{Reff}}$

- Received power (simplified):
$P_{Rx} = \left(\frac{P_T G_T}{4\pi R^2}\right)\left(\frac{\lambda^2}{4\pi}\right)G_R = (P_T G_T)\left(\frac{\lambda}{4\pi R}\right)^2 G_R = (P_T G_T)\left(\frac{1}{L_{FS}}\right)G_R$

- Friss Equation: $P_{Rx} = \left(\frac{P_{Tx}G_{Tx}G_{Rx}}{L_{Tx}L_{Rx}}\right)\left(\frac{\lambda}{4\pi d}\right)^2$

- System noise temperature at antenna output:
$T_1 = T_A + (L_{FRX} - 1)T_F + \frac{T_{\text{eff}}}{G_{RX}}$

- System noise temperature at receiver input: $T_{\text{sys}} = T_2 = \frac{T_1}{L_{FRX}} = \frac{T_A}{L_{FRX}} + T_F(1 - \frac{1}{L_{FRX}}) + T_{\text{eff}}$

- Amplifier noise temperature: $T_{\text{amplifier}} = T_{\text{eff}} = T_0(10^{\frac{NF}{10}} - 1)$

A.8 Chapter 9: Link Performance

- Signal (Carrier) to Noise Power Ratio:
$\frac{C}{N} = EIRP + L_{\text{FSPL}} + L_{\text{atm}} + \frac{G_R}{T_{\text{sys}}} - k - 10\log 10(B)$

- Carrier to Noise Power Ratio referenced to 1 Hz:
$\frac{C}{N_0} = EIRP + L_{\text{FSPL}} + L_{\text{atm}} + \frac{G_R}{T_{\text{sys}}} - k$

- Energy per bit to noise spectral density: $\frac{E_b}{N_0} = \frac{P_T L_{\text{feeder}} G_T L_{\text{FSPL}} L_{\text{atm}} G_R}{k T_{\text{sys}} R}$

- C/N (computed from E_b/N_0): $\frac{C}{N} = \frac{E_b}{N_0} \times \frac{R}{B}$

- Shannon's Limit: $R_{\max} = B\log_2\left(1 + \frac{C}{N}\right)$

Glossary

A | B | C | E | F | G | I | L | M | O | P | R | S | T | U | V

A

Apogee — the point in the orbit that is farthest from the Earth. 16, 20–27, 29, 37, 42, 50, 56, 57, 62

Argument of perigee — represented as ω; the angle from the line of nodes to the position vector at perigee. 33, 35, 36, 53, 141

Atmospheric drag — the drag force acting on satellites as a result of atmospheric friction; a mechanism of orbital decay. 69

Azimuth — the angle in the spherical coordinate system that is defined clockwise from due north to the satellite. 92

B

Beamwidth — the angular span of the main lobe of an antenna's beam. It describes how quickly gain decreases away from boresight, or the peak direction. 99–101, 121

C

Circular Orbit — an orbit with an eccentricity of zero, which suggests that the radius of apogee equals the radius of perigee. 21, 24, 37, 42, 43, 57, 58, 61–63

Communication link — the connection between a transmitter and a receiver for the primary purpose of data transmission. 73

E

Eccentric anomaly — the angle from the center of the orbit to a point that creates a right angle with the apse line, the line connecting the apogee and perigee, on an theoretical circle that includes the orbit. This theoretical circular orbit is used to simplify the calculation of the satellite's position. 37

Eccentricity — describes the shape of a conic section. Remember that circular orbits have an eccentricity of zero. 22–25, 29, 42, 50, 141

Elevation — the angle in the spherical coordinate system that is defined from the horizon to the satellite. 111–113, 115, 126

Ellipse — a curve that is the locus of all points in the plane for which the sum of distances r_1 and r_2 from two fixed points F_1 and F_2 (the foci) separated by a distance of $2c$ is a given positive constant $2a$. An ellipse may also be defined in terms of one focal point and a line outside the ellipse called the directrix. For all points on the ellipse, the ratio between the distance to the focus and the distance to the directrix is a constant. This constant ratio is called the eccentricity of the ellipse. 20, 21, 37, 41, 42, 103, 104

Elliptical Orbit — an orbit with an eccentricity between 0 and 1. 17, 21, 24, 38, 42, 43, 57

Equatorial orbit — an orbit with inclination of 0 degrees. 26

F

Foci — the plural version of the word focus. Every ellipse has two focus points. Foci always lay on the major axis, which happens to be the longest axis, equally spaced from the center. The foci help define the shape of the ellipse. 22, 28, 57

G

Gain — the ratio between the actual and isotropic power in the given direction of an antenna. Used to quantify how effectively power is transmitted or received in that direction. 80, 82, 86, 93–95, 97–100, 105, 106, 122–124, 127, 130, 141, 142

GEO — Geostationary Orbit has an altitude of 35786 km, and an inclination of 0 degrees (equatorial). 15, 16, 44, 64, 65, 70, 111, 121, 134, 135, 141

Graveyard orbit — an orbit for the purpose of post mission disposal that lies away from common operational orbits. Graveyard orbits can contain "ghost" satellites that are no longer operational. 69–71

I

Inclination — tilt of the orbital plane, measured as the angle between the Earth's equatorial and the orbital plane. Inclination determines the north and south latitude of a ground track. 15, 25, 26, 33, 36, 51, 53, 141

L

Law of Areas — Kepler's Second Law of Planetary Motion, states that the vector from the Sun to the planet sweeps equal areas in equal times. As a result, objects closer to the Sun travel faster than those farther away from the Sun. 41

LEO — Low Earth Orbit is defined as an orbit with altitude below around 2,000 km. 15, 17, 22, 25, 45, 51, 64, 65, 67–69, 71, 141

M

Mean anomaly — the fraction of an elliptical orbit's period that has elapsed since passing periapsis, expressed as an angle. It is derived from the eccentric anomaly, E, as in Equation 2.10. 37, 38, 54

Mean motion — the average rate of motion of the satellite; related to the gravitational parameter, μ, and the semi-major axis of an elliptical orbit, a, as in Equation 2.6. 34, 37, 53

MEO — Medium Earth Orbit lies between LEO and GEO. 15, 44, 69, 141

Molniya Orbit — a type of highly elliptical orbit with an eccentricity of 0.74, an inclination of 63.4 degrees (the critical inclination which has no apogee drift), and an orbital period of approximately half a sidereal day. This orbit provides coverage to extreme latitudes, and was originally designed by the Soviets to enable long dwell times over the northern hemisphere. 17

O

Orbit equation — the "orbit equation" defines the path of a satellite of mass, m_2, around the Earth of mass, m_1, relative to m_1. Keep in mind that h, μ, and e are constants. Theta is the true anomaly, also referred to as ν, and is the angle between the perigee and the spacecraft's position as referenced in Equation 3.17. 50

Orbital decay — over time, a satellite's orbit decays due to friction-like mechanisms which transfer energy from the orbital motion. For bodies in LEO, the most significant effect is atmospheric drag. 69

Orbital period — the time required for the Earth to complete one full rotation under the satellite. For circular orbits, speed is constant so the period is the circumference divided by the speed. For elliptical orbits, the equation takes a similar, but slightly different form as seen in Equation 3.2. 16, 29, 41, 43–45, 56, 141

P

Perigee — the point in the orbit that is nearest to the Earth. 17, 20–25, 27, 29, 32, 33, 37, 42, 46, 50, 56, 58, 61

Polar — when inclination, i, is 90 degrees. 26, 99

Polarization — the orientation of the electric field of a wave. 102–105, 142

Prograde — when inclination, i, is greater than 90 degrees. Use the right hand rule. Curl your fingers in the direction of the satellite orbit. If your thumb points upward, the orbit is prograde. 26, 27

R

RAAN — the angle between Vernal Equinox and the node line. The intersection of the equatorial and orbital plane is called the node line. The satellite crosses the node line at two points. The ascending node is where the satellite crosses the equatorial plane and is moving upwards. RAAN is the angle from the coordinate system x-axis to the line of nodes. The RAAN of your satellite's orbit is just the angle, measured at the center of the Earth, between the place the Sun's orbit ascends the equator, and the place your satellite's orbit ascends the equator. 31, 32, 35, 36, 53, 141

Re-entry — when a satellite's orbital altitude decreases such that it re-enters Earth's atmosphere. Satellites in LEO either actively perform maneuvers to lower their altitude and re-enter, or passively re-enter due to orbital decay from atmospheric drag. Ultimately, these spacecraft burn up in the Earth's atmosphere instead of moving to a graveyard orbit. 68, 69, 71

Retrograde — when inclination, i, is less than 90 degrees. Use the right hand rule. Curl your fingers in the direction of the satellite orbit. If your thumb points downward, the orbit is retrograde. 26, 27

S

Semi-major axis — the average of the radius to perigee and the radius to apogee. 20–22, 29, 34, 41–46, 50, 141

Sidereal day — approximately 23 hours, 56 minutes, and 4 seconds, or 23.93 hours, nearly 4 minutes shorter than a solar day. Astronomers use this system to locate celestial objects, because it is based on the "Earth's rate of rotation measured relative to the fixed stars (inertial space)". In other words, Earth makes one rotation (360 degrees) around its axis in a sidereal day. 16, 30

Sidereal time — measured by the rotation of the earth relative to the fixed stars. 16

Solar day — 24 hours, and is also the time it takes for the Sun to return to the same position over Earth. 16

SSO — nearly polar orbits in which the satellite precesses through one complete revolution per year, so it always maintains the same relationship with the Sun. Specifically, the satellite's orbital plane moves 1 degree eastward each day with respect to the Earth, keeping pace with the Earth's movement around the Sun. This places the satellite in constant sunlight, which allows the solar panels to be continuously illuminated. This is also useful for imaging satellites, because imaging in the visible range of the electromagnetic spectrum means you need daylight to see. 17, 141

T

True anomaly — represented as ν, f, or θ; the angle between perigee and the position of the spacecraft from the center of the Earth. Top down view of orbit shows the true anomaly more clearly. Satellites in elliptical orbits don't move at a constant speed! Therefore, the true anomaly doesn't change at a constant rate. 33, 34, 37, 38, 50, 141, 151

U

Universal Law of Gravitation — states that two bodies of mass m and M attract each other with a force that is proportional to their masses and that is inversely proportional to the square of the distance, r, between them as seen in Equation 3.1. 42

V

Vernal Equinox — the place where the sun rises on the first day of Spring, when the day and night have the same duration. This happens twice a year. 32

Bibliography

[1] George Biddell Airy. "Description of the Transit Circle of the Royal Observatory, Greenwich". In: *Greenwich Observations in Astronomy, Magnetism and Meteorology made at the Royal Observatory, Series 2* 29 (Jan. 1869), H1–HPXVI.

[2] E. F. W. Alexanderson. "Polarization of radio waves". In: *IEEE* 45.7 (1926), pp. 636–640.

[3] Roger R. Bate, Donald D. Mueller, and Jerry E. White. *Fundamentals of AstroDynamics*. Dover Publications, INC., 1971.

[4] Luc Blanchet. "On the two-body problem in general relativity". In: *arXiv* (2001).

[5] Tim W. C. Brown and Alex C. H. Tay. "On the Benefits of Polarization for Fixed Area Television White Space Devices". In: *IEEE* 62.3 (2014), pp. 1147–1156.

[6] Howard D. Curtis. "Chapter 5 - Preliminary Orbit Determination". In: *Orbital Mechanics for Engineering Students (Third Edition)*. Ed. by Howard D. Curtis. Third Edition. Boston: Butterworth-Heinemann, 2014, pp. 239–298. ISBN: 978-0-08-097747-8.

[7] Lance D. Davis. *March Equinox Brings 2 Seasons: Spring, Autumn*. NASA, 2021.

[8] Dave Doody. *Basics of Space Flight*. NASA, 2023. Chap. 4.1.

[9] European Space Agency. *Impact chip*. 2016.

[10] Federal Communications Commission. *FCC adopts new '5-year rule' for deorbiting satellites*. Sept. 2022.

[11] Francis Lyall. "The International Telecommunication Union: Origin and Role". In: *Communication* (2021).

[12] M. Garcia. *Space debris and human spacecraft*. NASA, Apr. 2015.

[13] David H. Hathaway. "The Solar Cycle". In: *Living Reviews in Solar Physics* 12.4 (2015).

[14]   Hans J. Haubold. "Kepler's laws". In: *Encyclopedia of Planetary Science* (1997), pp. 379–380.

[15]   Walter Hohmann. *Die Erreichbarkeit der Himmelskörper*. R. Oldenbourg Verlag, 1960.

[16]   International Telecommunications Union. *Radio Regulations*. 2023.

[17]   Russell JL. "Kepler's Laws of Planetary Motion: 1609–1666". In: *The British Journal for the History of Science* (1964).

[18]   Kessler, D. J. and Johnson, N. L. and Liou, J.-C. and Matney, M. "The Kessler syndrome: Implications to future Space Operations". In: *Advances in the Astronautical Sciences* 137 (2010).

[19]   Stanley Q. Kidder and Thomas H. Vonder Haar. "On the Use of Satellites in Molniya Orbits for Meteorological Observation of Middle and High Latitudes". In: *Journal of Atmospheric and Oceanic Technology* 7.3 (1990), pp. 517–522.

[20]   L. Laccourreye and O. Laccourreye. "A century ago, the birth of the decibel". In: *European Annals of Otorhinolaryngology, Head and Neck Diseases* (2024).

[21]   J.C. Liou. *The 2019 U.S. Government Orbital Debris Mitigation Standard Practices*. NASA, 2019.

[22]   Whitney Lohmeyer, Raichelle Aniceto, and Kerri Cahoy. "Communication satellite power amplifiers: current and future SSPA and TWTA technologies". In: *International Journal of Satellite Communications and Networking* 34.2 (2015), pp. 95–113.

[23]   Greg Markowsky. "A retelling of Newton's work on Kepler's Laws". In: *Expositiones Mathematicae* 29.3 (2011), pp. 253–282.

[24]   *Taking Garbage Outside: The Geostationary Orbit and Graveyard Orbits*. 57th International Astronautical Congress. 2006.

[25]   W. G. Perrin. "The Prime Meridian". In: *The Mariner's Mirror* 13.2 (1927), pp. 109–124.

[26]   Physics Today. *Guglielmo Marconi*. 2017.

[27]   Nesrin Sarigul-Klijn, Chris Noel, and Martinus M Sarigul-Klijn. "Air Launching Eart-to-Orbit Vehicles: Delta V gains from Launch Conditions and Vehicle Aerodynamics". In: *American Institute of Aeronautics and Astronautics* (2004).

[28]   Werner Schulz. "Walter Hohmann's contributions towards space flight: an appreciation on the occasion of the centenary of his birthday". In: *Acta Astronautica* 7.11 (1980), pp. 1213–1227.

[29] C.E. Shannon. "A Mathematical Theory of Communication". In: *Bell System Technical Journal* 27 (1948), pp. 379–423.

[30] Asif A. Siddiqi. *Beyond Earth: A Chronicle of Deep Space Exploration, 1958-2016*. NASA, 2018, pp. 11–12.

[31] B.G.A. Smith et al. "Ionospheric drag for accelerated deorbit from upper low earth orbit". In: *Acta Astronautica* 176 (2020), pp. 520–530.

[32] United Nations Office for Outer Space Affairs. *Register of Objects Launched into Outer Space*. 2024.

[33] David R. Williams. *Earth Fact Sheet*. NASA, 2024.

Printed in Great Britain
by Amazon